T0321018

Wide Bandgap Semiconductor Spintronics

Wide Bandgap Semiconductor Spintronics

Vladimir Litvinov

PAN STANFORD PUBLISHING

Published by

Pan Stanford Publishing Pte. Ltd.
Penthouse Level, Suntec Tower 3
8 Temasek Boulevard
Singapore 038988

Email: editorial@panstanford.com
Web: www.panstanford.com

British Library Cataloguing-in-Publication Data
A catalogue record for this book is available from the British Library.

Wide Bandgap Semiconductor Spintronics

ISBN 978-981-4669-70-2 (Hardcover)
ISBN 978-981-4669-71-9 (eBook)

Printed in the USA

To my wife, Valeria, and
my children, Natasha and Vlady

Contents

Preface

The field of spintronics is currently being explored in various directions. One of them, semiconductor spintronics, is of particular recent interest since materials developed for electronics and optoelectronics are gradually becoming available for spin-manipulation-related applications, e.g., spin-transistors and quantum logic devices allowing the integration of electronic and magnetic functionalities on a common semiconductor template.

The scope of this book is largely concerned with the spintronic properties of III-V Nitride semiconductors. As wide bandgap III-Nitride nanostructures are relatively new materials, particular attention is paid to the comparison between zinc-blende GaAs- and wurtzite GaN-based structures where the Rashba spin-orbit interaction plays a crucial role in voltage-controlled spin engineering.

The book also deals with topological insulators, a new class of materials that could deliver sizeable Rashba spin splitting in the surface electron spectrum when implemented into a gated device structure. Electrically driven zero-magnetic-field spin-splitting of surface electrons is discussed with respect to the specifics of electron-localized spin interaction and voltage-controlled ferromagnetism.

Semiconductor spintronics has been explored and actively discussed and various device implementations have been proposed along the way. Writings on this topic appear in the current literature. This book is focused on the materials science side of the question, providing a theoretical background for the most common concepts of spin-electron physics. The book is intended for graduate students and may serve as an introductory course in this specific field of solid state theory and applications. The book covers generic topics in spintronics without entering into device specifics since the overall goal of the enterprise is to give instructions to be used in solving problems of a general and specific nature.

Chapter 1 deals with the electron spectrum in bulk wurtzite GaN and the origin of linear terms in energy dispersion. Attention is paid to the symmetry and features of wurtzite spintronic materials which differentiate them from their cubic GaInAs-based counterparts.

Rashba and Dresselhaus spin-orbit terms in heterostructures with one-dimensional confinement are considered in Chapter 2, where typical spin textures are discussed in relation to in-plane electron momentum. This chapter also presents the microscopic derivation of the Rashba interaction in wurtzite quantum wells that allows electron spin-splitting to be related to the material and geometrical parameters of the structure. In particular, we discuss Rashba spin splitting in a structurally symmetric wurtzite quantum well to focus on the polarization-field induced Rashba interaction.

Vertical tunneling through a single barrier and a polarization-field distorted Al(In)GaN/GaN quantum well, as a possible spin-injection mechanism, is considered in Chapter 4.

Chapters 5 and 6 are devoted to a detailed theoretical description of mechanisms of ferromagnetism in magnetically doped semiconductors, specifically in the III-V Nitrides. These chapters discuss the indirect exchange interaction in metals of any dimension and in semiconductors. Emphasis is placed on the specific feature of the indirect exchange interaction in a one-dimensional metal. Also, the standard mean-field approach to ferromagnetic phase transition is described, as is the percolation picture of phase transition in certain systems, for example, wide bandgap semiconductors, for which mean-field theory breaks down.

The electronic properties of topological Bi_2Te_3 insulators are discussed in Chapter 7, where the semiconductor is taken as an example. Topological insulator film biased with a vertical voltage presents a system with voltage-controlled Rashba interaction and it is of interest in relation to possible spintronic applications. Surface electrons in the biased topological insulator are spin-split and this affects the indirect exchange interaction between magnetic atoms adsorbed onto a surface. The calculation of indirect exchange in a topological insulator is given in Chapter 8.

I would like to thank V. K. Dugaev, H. Morkoc, and D. Pavlidis for many useful discussions of the topics discussed in this book and Toni Quintana for carefully reading and correcting the text.

Chapter 1

GaN Band Structure

To deal with the spin and electronic properties of wurtzite III-nitride semiconductors and understand the specific features that differentiate them from zinc blende III-V materials, one has to know the energy spectrum. The energy spectrum gives us all necessary information about how electron spin is related to its momentum; and that is the key information we need in order to use the material in various spintronic applications.

1.1 Symmetry

Ga(Al,In)N crystallizes in two modifications: zinc blende and wurtzite. The crystal structure of the wurtzite GaN belongs to the space group $P6_3mc$ (International notation) or C_{6v}^4 (Schönflies notation). The unit cell is shown in Fig. 1.1.

In the periodic lattice potential, the electron Hamiltonian is invariant to lattice translations, so the wave function should be an eigenfunction $\psi(\mathbf{r})$ of the translation operator:

$$\psi(\mathbf{r} + \mathbf{R}) = \psi(\mathbf{r})\exp(i\mathbf{kR}), \tag{1.1}$$

where $\exp(i\mathbf{kR})$ is the eigenvalue of the translation operator, and $\mathbf{R}$ is the arbitrary lattice translation. This condition is the

Wide Bandgap Semiconductor Spintronics
Vladimir Litvinov

ISBN 978-981-4669-70-2 (Hardcover), 978-981-4669-71-9 (eBook)
www.panstanford.com

consequence of symmetry only and it presents a definition of the wave vector. In an infinite crystal, the wave vector would be a continuous variable. Since we are dealing with a crystal of finite size, we have to impose boundary conditions on the wave function. This can be done in two ways. First, we may equate the wave function to zero outside the boundaries of the crystal. This would correspond to taking the surface effects into account. If we are not interested in finite-size (or surface) effects, there is a second option: We assume that the crystal comprises an infinite number of the periodically repeated parts of volume V (volume of a crystal) and then impose the Born–von Karman cyclic boundary conditions:

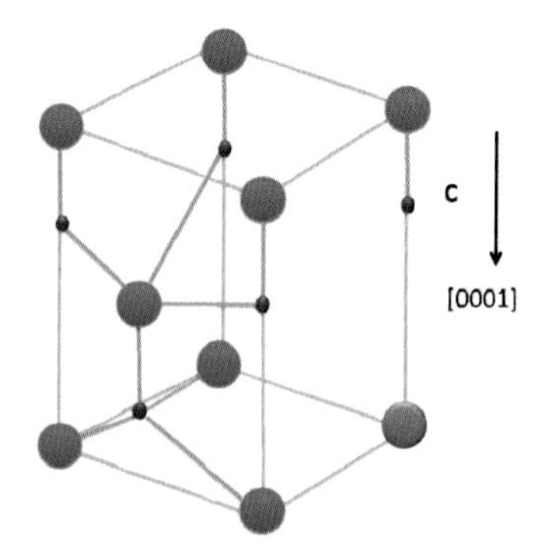

Figure 1.1 Unit cell of a GaN crystal. Large spheres represent Ga sites.

$$\psi(\mathbf{r} + N_i \mathbf{b}_i) = \psi(\mathbf{r}), \tag{1.2}$$

where $\mathbf{b}_i$ are basis vectors of the Bravais lattice. From Eqs. (1.1) and (1.2)

$$\exp(i\mathbf{k}\mathbf{L}_i) = 1$$

$$\mathbf{k}_i = \frac{2\pi}{L_i} m_i,\ m_i = 0, \pm 1, \ldots, \tag{1.3}$$

where $\mathbf{L}_i = \mathbf{b}_i N_i$ is the linear size of the crystal of volume V in the direction $\mathbf{b}_i$. Thus, the wave vector takes discrete values, so all integrals over the wave vectors that may appear in the theory should be replaced by summation over the discrete variable $\mathbf{k}$.

In bulk materials ($V \to \infty$), the wave vector is a quasi-continuous variable. The exact summation over $\mathbf{k}$ can be replaced by the integral

$$\sum_{\mathbf{k}} f(\mathbf{k}) \to \frac{V}{(2\pi)^3} \int f(\mathbf{k}) d\mathbf{k}. \tag{1.4}$$

The wave vector can be handled in much the same way as it was in a free space. However, the difference between **k** in free space and in the periodic lattice field is that the lattice periodicity introduces an ambiguity to the wave vector; that is it is defined up to the reciprocal lattice vector **K**. Formally, the reciprocal space is defined by expanding arbitrary lattice periodic function into the Fourier series:

$$\zeta(\mathbf{r}) = \sum_{\mathbf{K}} \zeta(\mathbf{K}) \exp(i\mathbf{K}\mathbf{r}) \tag{1.5}$$

Let's expand $\zeta(\mathbf{r})$ displaced on the lattice vector **R**:

$$\zeta(\mathbf{r} + \mathbf{R}) = \sum_{\mathbf{K}} \zeta(\mathbf{K}) \exp(i\mathbf{K}(\mathbf{r} + \mathbf{R})) \tag{1.6}$$

As the displacement **R** cannot change $\zeta(\mathbf{r})$ due to the lattice periodicity, the condition $\zeta(\mathbf{r}) = \zeta(\mathbf{r} + \mathbf{R})$ holds

$$\sum_{\mathbf{K}} \zeta(\mathbf{K}) \exp(i\mathbf{K}\mathbf{r}) = \sum_{\mathbf{K}} \zeta(\mathbf{K}) \exp(i\mathbf{K}(\mathbf{r} + \mathbf{R})). \tag{1.7}$$

It follows from Eq. (1.7)

$$\exp(i\mathbf{K}\mathbf{R}) = 1. \tag{1.8}$$

Equation (1.8) defines the reciprocal lattice (or dual lattice) for vectors **K**. From Eqs. (1.1) and (1.8), we conclude that replacement $\mathbf{k} \to \mathbf{k} + \mathbf{K}$ does not change the wave function, Eq. (1.1), so **k** and $\mathbf{k} + \mathbf{K}$ are equivalent. This means that the electron kinematics in the lattice can be fully described by the wave vectors in the finite part of the reciprocal space, the first Brillouin zone (BZ). BZ for the crystals of the wurtzite family is illustrated in Fig. 1.2.

The wave function that satisfies Eq. (1.1) is the Bloch function

$$\psi_{n\mathbf{k}}(\mathbf{r}) = \frac{1}{\sqrt{V}} u_{n\mathbf{k}}(\mathbf{r}) \exp(i\mathbf{k}\mathbf{r}), \tag{1.9}$$

which is the modulated plane wave normalized on the crystal volume V, $u_{n\mathbf{k}}(\mathbf{r})$ is the lattice periodic Bloch amplitude, and n is the band index.

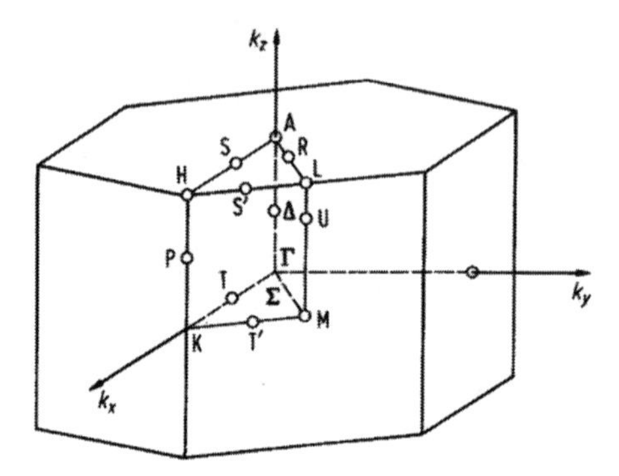

Figure 1.2 First Brillouin zone for wurtzite crystal. Capital letters indicate the high symmetry points of wave vector $\mathbf{k}$ in BZ.

Symmetry dictates that the Hamiltonian is invariant to all transformations of the space group C_{6v}^4. When an element of the space symmetry group acts on a crystal, it transforms both the space coordinate $\mathbf{r}$ and the wave vector $\mathbf{k}$. In each high symmetry point shown by capital letters in Fig. 1.2, there exists a point subgroup of rotations that leave the corresponding wave vector unchanged. Both this subgroup and the time reversal symmetry determine the energy spectrum near the $\mathbf{k}$-point of high symmetry. In III-nitrides the energy spectrum that is responsible for the electrical, magnetic, and optical properties of the material, lies near the point Γ ($k = 0$) The relevant energy levels for spin-up and spin-down electrons include six valence and two conduction bands each corresponding to irreducible representations of the point group C_{6v} [1, 2]. Below the spectrum at the Γ-point will be constructed using Luttinger–Kohn basis wave functions.

1.2 Hamiltonian

Energy bands can be found as eigenvalues of the Schrödinger equation:

$$H\psi = E\psi$$

$$H = \frac{p^2}{2m_0} + V(\mathbf{r}) + \frac{\hbar}{4m_0^2c^2}\boldsymbol{\sigma} \times \boldsymbol{\nabla}V(\mathbf{r})\cdot \mathrm{p}, \tag{1.10}$$

where $\sigma_{x,y,z}$ are the Pauli matrices; e, m_0 are the free electron charge and mass, respectively; $\mathbf{p} = -i\hbar\boldsymbol{\nabla}$ is the momentum operator; and $V(\mathbf{r})$ is the electron potential energy in the periodic crystal field. The third term in Eq. (1.10) represents the spin–orbit interaction.

Using $\psi_{n\mathbf{k}}(\mathbf{r})$ from Eq. (1.9) we obtain the equation for Bloch amplitudes:

$$H_{\mathbf{k}}u_{n\mathbf{k}} = E_{n\mathbf{k}}u_{n\mathbf{k}},$$

$$H_{\mathbf{k}} = H_0 + \frac{\hbar^2 k^2}{2m_0} + \frac{\hbar\mathbf{kp}}{m_0} + H_1 + H_2,$$

$$H_0 = \frac{p^2}{2m_0} + V(\mathbf{r}),\ H_1 = \frac{\hbar}{4m_0^2c^2}\boldsymbol{\sigma}\cdot\boldsymbol{\nabla}V(\mathbf{r})\times\mathbf{p},\ H_2 = \frac{\hbar^2}{4m_0c^2}\mathbf{k}\cdot\boldsymbol{\sigma}\times\boldsymbol{\nabla}V(\mathbf{r}). \tag{1.11}$$

In order to find the eigenvalues $E_{n\mathbf{k}}$, one has to choose the full set of known orthogonal functions that create the initial basis on which we can expand the unknown amplitudes $u_{n\mathbf{k}}(\mathbf{r})$. As we are looking for the spectrum in the vicinity of the Γ-point, the set of band edge Bloch amplitudes $u_{n0}(\mathbf{r})$ can serve as the basis wave functions (Luttinger–Kohn representation). Within *kp*-perturbation theory, the third, fourth, and fifth terms in the Hamiltonian (1.11) are being treated as a perturbation.

The Hamiltonian H_0 does not include the spin–orbit interaction, so we restrict our consideration to the three band edge energy levels that correspond to the irreducible representations of the point group C_{6v}: the conduction band Γ_{1c}, the double degenerate in orbital momentum valence band Γ_6, and one more valence band Γ_1. Relevant bands are shown in Fig. 1.3a. The levels are degenerate and the degeneracy is shown in parentheses.

With account for the spin variable, the basis comprises eight bands: three valence bands and one conduction band. The Bloch amplitude can be represented as a linear combination

$$u_{\mathbf{k}}(\mathbf{r}) = \frac{1}{\sqrt{\Omega}}\sum_{n=1}^{8} C_{n\mathbf{k}}u_{n0}(\mathbf{r}), \tag{1.12}$$

where the amplitudes $u_{n0}(\mathbf{r})$ are orthogonal and normalized on a unit cell volume Ω:

$$\int u_{n0}^*(\mathbf{r})u_{n'0}(\mathbf{r})d\Omega \equiv \langle u_{n0}(\mathbf{r})|u_{n'0}(\mathbf{r})\rangle = \delta_{nn} \tag{1.13}$$

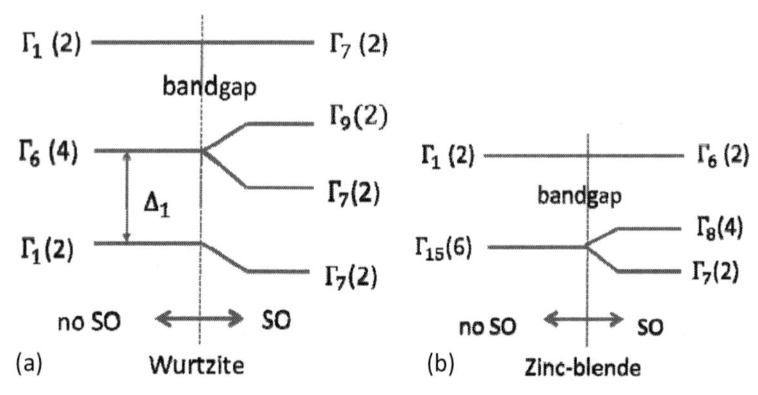

Figure 1.3 Structure of the Γ-bands in wurtzite (a) and zinc blende (b) crystals. The left hand side of each panel shows the basis states with no spin–orbit interaction taken into account; Δ_1 is the crystal-field energy splitting.

Conduction and valence bands in GaN stem from *s*- and *p*-orbitals of Ga and N. The conduction band has spherical *s*-symmetry, so the Bloch amplitude at the band edge can be chosen to be a spherical *s*-orbital, Y_{00}. The three valence bands obey the *p*-symmetry and can be chosen as linear combinations of spherical harmonics that are the eigenfunctions of $\hat{L}_z$, *z*-component of the orbital momentum with $L = 1$ (*p*-state). Once we choose the principal axis *Z* along the *c*-direction (see Fig. 1.1) the basis spherical harmonics can be expressed in terms of *p*-orbitals $|X>$, $|Y>$, $|Z>$ shown in Fig. 1.4. The operator $\hat{L}_z$ has three eigenvalues $l = -1, 0, 1$ and the corresponding spherical harmonics have the form:

$$Y_{11} = -\frac{1}{\sqrt{2}}|(X+iY)>,\ Y_{1-1} = \frac{1}{\sqrt{2}}|(X-iY)>,\ Y_{10} = |Z>. \tag{1.14}$$

The orbital momentum operator is defined as

$$\hat{\mathbf{L}} = [\mathbf{r}\times\mathbf{p}] = -i\hbar[\mathbf{r}\times\nabla]. \tag{1.15}$$

It is straightforward to check that the set (1.14) comprises $2l + 1 = 3$ eigenfunctions of $\hat{L}_z$ with corresponding eigenvalues $-1, 0, 1$:

$$-i\left(x\frac{\partial}{\partial y}-y\frac{\partial}{\partial x}\right)|(X+iY)>=|(X+iY)>,$$
$$-i\left(x\frac{\partial}{\partial y}-y\frac{\partial}{\partial x}\right)|(X-iY)>=-|(X+iY)>,$$
$$-i\left(x\frac{\partial}{\partial y}-y\frac{\partial}{\partial x}\right)|Z>=0. \quad (1.16)$$

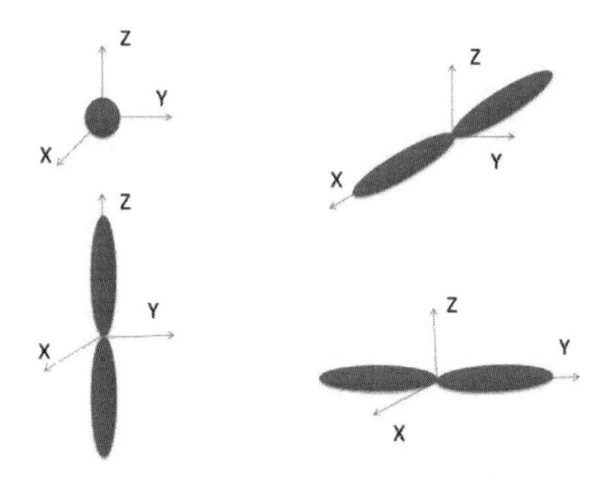

Figure 1.4 Electron density in *s*- and *p*-orbitals, $|S>, |X>, |Y>, |Z>$.

Finally, not accounting for spin–orbit interaction, the basis set can be written as

$$\Gamma_{1c}(2): u_1 = |iS\uparrow>, u_5 = |iS\downarrow>,$$
$$\Gamma_6(2): u_2 = -\frac{1}{\sqrt{2}}|(X+iY)\uparrow>, u_6 = \frac{1}{\sqrt{2}}|(X-iY)\downarrow>,$$
$$\Gamma_6(2): u_3 = \frac{1}{\sqrt{2}}|(X-iY)\uparrow>, u_7 = -\frac{1}{\sqrt{2}}|(X+iY)\downarrow>,$$
$$\Gamma_{1v}(2): u_4 = |Z\uparrow>; u_8 = |Z\downarrow>, \quad (1.17)$$

where $|S\uparrow> = |Y_{00}| * |\uparrow>$, and $|\uparrow> = \begin{pmatrix}1\\0\end{pmatrix}$, $|\downarrow> = \begin{pmatrix}0\\1\end{pmatrix}$ are the spinor wave functions that correspond to spin-up and down states, respectively. Equation (1.17) forms the "quasi-cubic approximation" that neglects the anisotropy of wurtzite structure and represents basis wave functions in the same way as in zinc blende materials with the axis Z parallel to the [111] direction. The validity of the "quasi-cubic approximation" will be discussed below.

In a first-order kp-approximation, the Hamiltonian matrix can be written by making use of (1.17):

$$H_{nn'}(\mathbf{k}) = \int < u_n(\mathbf{r})|H_{\mathbf{k}}|u_{n'}(\mathbf{r}) > d\Omega \tag{1.18}$$

Some of the matrix elements are equal to zero as a result of the symmetry of the basis functions:

$$H = \begin{bmatrix} E_c & -\frac{P_2k_+}{\sqrt{2}} & \frac{P_2k_-}{\sqrt{2}} & P_1k_z & 0 & 0 & 0 & 0 \\ -\frac{P_2k_-}{\sqrt{2}} & F & 0 & 0 & 0 & 0 & 0 & 0 \\ \frac{P_2k_+}{\sqrt{2}} & 0 & G & 0 & 0 & 0 & 0 & \sqrt{2}\Delta_3 \\ P_1k_z & 0 & 0 & \lambda & 0 & 0 & \sqrt{2}\Delta_3 & 0 \\ 0 & 0 & 0 & 0 & E_c & \frac{P_2k_-}{\sqrt{2}} & -\frac{P_2k_+}{\sqrt{2}} & P_1k_z \\ 0 & 0 & 0 & 0 & \frac{P_2k_+}{\sqrt{2}} & F & 0 & 0 \\ 0 & 0 & 0 & \sqrt{2}\Delta_3 & -\frac{P_2k_-}{\sqrt{2}} & 0 & G & 0 \\ 0 & 0 & \sqrt{2}\Delta_3 & 0 & P_1k_z & 0 & 0 & \lambda \end{bmatrix}, \tag{1.19}$$

where

$$E_c = <iS\left|\frac{p^2}{2m_0} + V\right|iS>,$$

$$\Delta_1 = <X\left|\frac{p^2}{2m_0} + V\right|X> = <Y\left|\frac{p^2}{2m_0} + V\right|Y>,$$

$$\Delta_2 = \frac{\hbar^2}{8m_0^2c^2}\left(<Y\left|\nabla_y V\frac{\partial}{\partial x}\right|X> + <X\left|\nabla_x V\frac{\partial}{\partial y}\right|Y>\right),$$

$$\Delta_3 = \frac{\hbar^2}{4m_0^2c^2}<X\left|\nabla_x V\frac{\partial}{\partial z}\right|Z> = \frac{\hbar^2}{4m_0^2c^2}<Y\left|\nabla_y V\frac{\partial}{\partial z}\right|Z>, \tag{1.20}$$

and $k_{\pm} = k_x \pm ik_y \equiv k\exp(\pm i\varphi)$, $k^2 = k_x^2 + k_y^2$, $\tan(\varphi) = k_y/k_x$, $F = \Delta_1 + \Delta_2$, $G = \Delta_1 - \Delta_2$, $P_1 = \frac{\hbar^2}{m_0\sqrt{2}} < iS\left|\frac{\partial}{i\partial z}\right|Z>$, $P_2 = \frac{\hbar^2}{m_0\sqrt{2}} < iS\left|\frac{\partial}{i\partial x}\right|X> = \frac{\hbar^2}{m_0\sqrt{2}} < iS\left|\frac{\partial}{i\partial y}\right|Y>$, are momentum matrix elements, E_c is the position of the conduction band edge, Δ_1 and $\Delta_{2,3}$ are the parameters of the crystal field and spin–orbit interactions, respectively, $\lambda = E_{v0} = 0$ is the reference energy which would be the valence band edge position if the crystal-field and spin–orbit splitting were not taken into account.

Hexagonal symmetry makes the Hamiltonian (1.10) invariant with respect to $\beta = \pi/3$ rotation in the basal xy plane:

$$\begin{aligned} x' &= x\cos\beta + y\sin\beta \\ y' &= y\cos\beta - x\sin\beta \end{aligned} \tag{1.21}$$

That results in conditions

$$\begin{aligned} &<Y|H_0|Y> = <Y'|H_0|Y'> = \frac{3}{4} <X|H_0|X> + \frac{1}{4}<Y|H_0|Y>, \\ &<Y|H_0|Y> = <X|H_0|X>. \end{aligned} \tag{1.22}$$

Eigenvalues of the Hamiltonian found with $Det[\mathrm{H} - \mathrm{E}] = 0$ present the energy levels in Γ-point ($\mathbf{k} = 0$). Notations used below correspond to those in Fig. 1.3:

$$\begin{aligned} &\text{Conduction band, } \Gamma_{7c}: && E_c = E_g + \Delta_1 + \Delta_2, \\ &\text{Heavy holes band, } \Gamma_{9v}: && E_{v1} = \Delta_1 + \Delta_2, \\ &\text{Light holes band, } \Gamma_{7v}: && E_{v2} = \frac{\Delta_1 - \Delta_2}{2} + \sqrt{\left(\frac{\Delta_1 - \Delta_2}{2}\right)^2 + 2\Delta_3^2}, \\ &\text{Crystal-field split-off band, } \Gamma_{7v}: && E_{v3} = \frac{\Delta_1 - \Delta_2}{2} - \sqrt{\left(\frac{\Delta_1 + \Delta_2}{2}\right)^2 + 2\Delta_3^2}, \end{aligned} \tag{1.23}$$

The reference energy $E = 0$ corresponds to the lowest valence band E_{v3} when spin–orbit interaction is absent. The line-up of the valence bands in GaN, given in increasing order of their distance to the conduction band are heavy holes, light holes, and split-off band. This order is different in AlN, where the positive value of crystal-field splitting $\Delta_1 = 38$ meV [3] in GaN changes to a large negative value of −219 meV in AlN. As a result, in AlN heavy and light hole bands trade positions and the order becomes: light holes, heavy holes, split-off band, and the position of the conduction band is $E_c = E_{v2} + E_g$.

At finite **k** matrix (1.19) becomes more populated, so it is convenient to use the matrix identity that operates with blocks of lower dimensions and helps in finding eigenvalues:

$$Det\begin{bmatrix} A & B \\ C & D \end{bmatrix} = Det[AD - ACA^{-1}B] \tag{1.24}$$

Using (1.24), we rewrite $Det[H - \varepsilon] = 0$ and find energy bands $E(\mathbf{k})$:

$$\begin{aligned} &F(\varepsilon, k_z, k) = 0, \\ &F(\varepsilon, k_z, k) = [(\Delta_1 - \varepsilon)^2 - \Delta_2^2][(E_c - \varepsilon)E + P_1^2 k_z^2] - \varepsilon(\Delta_1 - \varepsilon)P_2^2 k^2 \\ &\qquad + \Delta_3^2[2(E_c - \varepsilon)(\Delta_1 + \Delta_2 - \varepsilon) - P_2^2 k^2], \\ &E(\mathbf{k}) = \varepsilon(\mathbf{k}) + \hbar^2 k^2/2m_0. \end{aligned} \tag{1.25}$$

Four solutions to Eq. (1.25) are a conduction and three valence bands each double spin degenerate.

The initial k-dependence of the energy level is determined by the heavy bare electron mass, so the levels are almost dispersionless. Additional dispersion from $\varepsilon(\mathbf{k})$ renormalizes the bare mass and imparts an effective mass which appears to be the result of coupling between the level and all other levels under consideration. Let us find effective masses of conduction electrons at $\varepsilon \to E_c = \Delta_1 + \Delta_2 + E_g$. An exact solution to Eq. (1.25) is not needed as the inverse effective mass can be found as the coefficient in the k^2 expansion of the exact energy:

$$\begin{aligned} &\frac{\hbar^2}{2}\left(\frac{1}{m_{cz}} - \frac{1}{m_0}\right) = -\left(\frac{\partial F}{\partial k_z^2}\right)\left(\frac{\partial F}{\partial \varepsilon}\right)^{-1}\Bigg|_{\substack{k\to 0 \\ \varepsilon \to E_c}} = \frac{(2\Delta_2 + E_g)}{(2\Delta_2 + E_g)(\Delta_1 + \Delta_2 + E_g) - 2\Delta_3^2} P_1^2, \\ &\frac{\hbar^2}{2}\left(\frac{1}{m_c} - \frac{1}{m_0}\right) = -\left(\frac{\partial F}{\partial k^2}\right)\left(\frac{\partial F}{\partial \varepsilon}\right)^{-1}\Bigg|_{\substack{k\to 0 \\ \varepsilon \to E_c}} = \frac{(\Delta_2 + E_g)(\Delta_1 + \Delta_2 + E_g) - \Delta_3^2}{E_g[(2\Delta_2 + E_g)(\Delta_1 + \Delta_2 + E_g) - 2\Delta_3^2]} P_2^2 \end{aligned} \tag{1.26}$$

Effective masses in the valence band cannot be described by Hamiltonian (1.19) as eight basis functions (1.17) do not form the full set: All other energy levels are neglected. However, remote bands cannot be neglected as they generate k_i^2 corrections and contribute to effective masses in all bands. In order to account for remote levels the Löwdin method is normally used so that the full set of basis functions is divided into two subsets: Subset A includes energy levels located close to the Fermi energy, and subset B includes more distant levels with the energy much larger than the actual energy of carriers contributing to the electrical and magnetic properties of the material:

$$u_{\mathbf{k}}(\mathbf{r}) = \sum_{n \in A} C_{n\mathbf{k}}\, u_{n0}(\mathbf{r}) + \sum_{j \in B} C_{j\mathbf{k}} u_{j0}(\mathbf{r}) \tag{1.27}$$

In the basis (1.27) the matrix Hamiltonian can be generally written as

$$H = \begin{bmatrix} A & C \\ C^{+} & B \end{bmatrix}, \tag{1.28}$$

where square blocks A and B correspond to subsets A and B, respectively, and the rectangular matrix C describes the coupling between A and B, so it comprises matrix elements of perturbation $\frac{\hbar \mathbf{kp}}{m_0} + H_1 + H_2$ on basis wave functions that belong to different sets. The Schrödinger equation with the Hamiltonian (1.28) can be written as

$$\begin{bmatrix} A & C \\ C^{+} & B \end{bmatrix} \begin{pmatrix} f \\ R \end{pmatrix} = E \begin{pmatrix} f \\ R \end{pmatrix}, \tag{1.29}$$

where the spinor wave function $\begin{pmatrix} f \\ R \end{pmatrix}$ consists of sub-spinors f and R each carrying a number of components equal to dimensions of matrices A and B, respectively. Notation C^{+} stands for the Hermitian conjugate (conjugate and transposed) matrix. Matrix Eq. (1. 29) can be written as a system of two equations

$$\begin{cases} Af + CR = Ef \\ C^{+}f + BR = ER \end{cases} \tag{1.30}$$

Solving the lower equation for R and substituting it into the upper one, we find the Schrödinger equation for A-block wave functions

$$Af + C(E-B)^{-1}C^{+}f = Ef. \tag{1.31}$$

So, the effective Hamiltonian for A-block takes the form

$$H = A + C(E-B)^{-1}C^{+} \tag{1.32}$$

The second term in Eq. (1.32) gives a correction originated from distant bands. This term is small if the energy distance between the level from subset A and remote bands is large. The Hamiltonian (1.32) is never used in its exact form. Instead, it is taken into account approximately, with desired accuracy on wave vector components. Energy corrections from remote bands enter in diagonal elements of A-matrix, they are proportional to k_i^2 and contribute to effective masses. With k_i^2 accuracy the Hamiltonian matrix has the form

$$H - \frac{\hbar^2 k^2}{2m_0} = \begin{bmatrix} E_c & -\frac{P_2k_+}{\sqrt{2}} & \frac{P_2k_-}{\sqrt{2}} & P_1k_z & 0 & 0 & 0 & 0 \\ -\frac{P_2k_-}{\sqrt{2}} & F & -K^* & -H^* & 0 & 0 & 0 & 0 \\ \frac{P_2k_+}{\sqrt{2}} & -K & G & H & 0 & 0 & 0 & \sqrt{2}\Delta_3 \\ P_1k_z & -H & H^* & \lambda & 0 & 0 & \sqrt{2}\Delta_3 & 0 \\ 0 & 0 & 0 & 0 & E_c & \frac{P_2k_-}{\sqrt{2}} & -\frac{P_2k_+}{\sqrt{2}} & P_1k_z \\ 0 & 0 & 0 & 0 & \frac{P_2k_+}{\sqrt{2}} & F & -K & H \\ 0 & 0 & 0 & \sqrt{2}\Delta_3 & -\frac{P_2k_-}{\sqrt{2}} & -K^* & G & -H^* \\ 0 & 0 & \sqrt{2}\Delta_3 & 0 & P_1k_z & H^* & -H & \lambda \end{bmatrix}, \tag{1.33}$$

where $D_{1,2,3,4}$ are the deformation potentials, ε_{ii} are the strain components, and coefficients A_{1-6} stem from remote bands corrections to the four bands under consideration [1, 4]:

$$F = \Delta_1 + \Delta_2 + \lambda + \theta + S_1 + S_2,\ G = \Delta_1 - \Delta_2 + \lambda + \theta + S_1 + S_2,$$
$$\lambda = S_1 + \frac{\hbar^2}{2m_0}[A_1 k_z^2 + A_2 k^2],\ \theta = \frac{\hbar^2}{2m_0}[A_3 k_z^2 + A_4 k^2],$$
$$K = \frac{\hbar^2}{2m_0} A_5 k_+^2 \equiv K_t \exp(2i\varphi),\ H = \frac{\hbar^2}{2m_0} A_6 k_+ k_z \equiv H_t \exp(i\varphi),$$
$$S_1 = D_1\varepsilon_{zz} + D_2(\varepsilon_{xx} + \varepsilon_{yy});\ S_2 = D_3\varepsilon_{zz} + D_4(\varepsilon_{xx} + \varepsilon_{yy}), \quad (1.34)$$

Hamiltonian (1.33) is non-diagonal even at $k = 0$. This means that basis wave functions (1.17) are not egenfunctions of (1.33), so they do not represent real Bloch amplitudes at the band edges. Real Bloch amplitudes are needed to calculate observables like energy shifts and transition amplitudes under external fields. In the Section 1.3, we transform the valence band Hamiltonian in order to find band edge Bloch amplitudes.

1.3 Valence Band Structure

Here we consider the Hamiltonian for the valence bands neglecting the coupling to the conduction band. This implies that the coupling between bare conduction and valence bands, induced by matrix elements H_{12}, H_{13}, H_{14} in (1.33), is taken into account in renormalized conduction band effective masses (1.26) and then the conduction band is decoupled from the valence bands. Thus, the Hamiltonian for the valence band is the 6 × 6 submatrix of (1.33) which can be written in the order of basis wave functions from thc set (1.17), $v_1, v_2, v_3, v_4, v_5, v_6$:

$$v_1 = -\frac{1}{\sqrt{2}}|(X + iY)\uparrow>,\ v_2 = \frac{1}{\sqrt{2}}|(X - iY)\uparrow>,\ v_3 = |Z\uparrow>,$$
$$v_4 = \frac{1}{\sqrt{2}}|(X - iY)\downarrow>,\ v_5 = -\frac{1}{\sqrt{2}}|(X + iY)\downarrow>,\ v_6 = |Z\downarrow>,$$
$$H = \begin{bmatrix} F & -K^* & -H^* & 0 & 0 & 0 \\ -K & G & H & 0 & 0 & \Delta_3\sqrt{2} \\ -H & H^* & \lambda & 0 & \Delta_3\sqrt{2} & 0 \\ 0 & 0 & 0 & H & -K & H \\ 0 & 0 & \Delta_3\sqrt{2} & -K^* & \lambda & -H^* \\ 0 & \Delta_3\sqrt{2} & 0 & H^* & -H & \lambda \end{bmatrix}, \quad (1.35)$$

The valence band Hamiltonian (1.35) is non-diagonal and its diagonalization would give us observable band dispersion in all three valence bands and eigenfunctions, which can be further used for the calculation of various observables, for instance, optical transition amplitudes, etc. There is a unitary transformation that block-diagonalizes the zinc blende [5, 6] and wurtzite [4] valence bands. The unitary transformation has the form:

$$T_1 = \begin{bmatrix} \alpha^* & 0 & 0 & \alpha & 0 & 0 \\ 0 & \beta & 0 & 0 & \beta^* & 0 \\ 0 & 0 & \beta^* & 0 & 0 & \beta \\ \alpha^* & 0 & 0 & -\alpha & 0 & 0 \\ 0 & \beta & 0 & 0 & -\beta^* & 0 \\ 0 & 0 & -\beta^* & 0 & 0 & \beta \end{bmatrix},$$

$$\alpha = \frac{1}{\sqrt{2}} \exp\left[i\left(\frac{3\pi}{2} + \frac{3\varphi}{2}\right)\right], \quad \beta = \frac{1}{\sqrt{2}} \exp\left[i\left(\frac{\pi}{2} + \frac{\varphi}{2}\right)\right], \qquad (1.36)$$

Transformation T_1 mixes basis functions (1.35) as

$$v' = T_1 \begin{pmatrix} v_1 \\ v_2 \\ v_3 \\ v_4 \\ v_5 \\ v_6 \end{pmatrix} = \begin{pmatrix} \alpha^* v_1 + \alpha v_4 \\ \beta v_2 + \beta^* v_5 \\ \beta^* v_3 + \beta v_6 \\ \alpha^* v_1 - \alpha v_4 \\ \beta v_2 - \beta^* v_5 \\ -\beta^* v_3 + \beta v_6 \end{pmatrix}, \qquad (1.37)$$

The goal of transformation is to remove spin-flip matrix elements from (1.35). As a result, in v'-representation, the Hamiltonian (1.35) is block-diagonalized into 3 × 3 spin-up and spin-down blocks as follows:

$$\widetilde{H} = T_1 H T_1^+ = \begin{bmatrix} F & K_t & -iH_t & 0 & 0 & 0 \\ K_t & G & \Delta_3\sqrt{2} - iH_t & 0 & 0 & 0 \\ iH_t & \Delta_3\sqrt{2} + iH_t & \lambda & 0 & 0 & 0 \\ 0 & 0 & 0 & F & K_t & iH_t \\ 0 & 0 & 0 & K_t & G & \Delta_3\sqrt{2} + iH_t \\ 0 & 0 & 0 & -iH_t & \Delta_3\sqrt{2} - iH_t & \lambda \end{bmatrix}$$

(1.38)

Upper and lower blocks relate to each other as $H^L = (H^U)^*$ and correspond to three bands of valence electrons of spin up and down, respectively. So, the spins become decoupled; however, the blocks remain non-diagonal even at $k = 0$ as they still contain spin–orbit terms $\Delta_3\sqrt{2}$ that mix light hole and crystal split bands.

In order to remove off-diagonal spin–orbit terms, we have to make another transformation that diagonalizes the submatrix

$$h = \begin{bmatrix} G & \Delta_3\sqrt{2} \\ \Delta_3\sqrt{2} & \lambda \end{bmatrix} \tag{1.39}$$

in each block. The full basis set transforms as

$$\begin{pmatrix} \Phi_1 \\ \Phi_2 \\ \Phi_3 \\ \Phi_4 \\ \Phi_5 \\ \Phi_6 \end{pmatrix} = T_2 \begin{pmatrix} v_1' \\ v_2' \\ v_3' \\ v_4' \\ v_5' \\ v_6' \end{pmatrix},$$

$$T_2 = \begin{bmatrix} 1 & 0 & 0 & 0 & 0 & 0 \\ 0 & b_+ & b_- & 0 & 0 & 0 \\ 0 & -b_- & b_+ & 0 & 0 & 0 \\ 0 & 0 & 0 & 1 & 0 & 0 \\ 0 & 0 & 0 & 0 & b_+ & b_- \\ 0 & 0 & 0 & 0 & -b_- & b_+ \end{bmatrix}, \quad b_+^2 + b_-^2 = 1$$

Off-diagonal terms $\Delta_3\sqrt{2}$ disappear from matrix (1.39) if

$$b_\pm = \frac{1}{\sqrt{2}}\left(1 \pm \frac{G-\lambda}{\sqrt{(G-\lambda)^2 + 8\Delta_3^2}}\right)^{1/2}. \tag{1.40}$$

For the upper block, for instance, the new basis has the form

$$\begin{aligned} \Phi_1 &= \alpha^* v_1 + \alpha v_4, \\ \Phi_2 &= (\beta v_2 + \beta^* v_5)\, b_+ + (\beta^* v_3 + \beta v_6) b_-, \\ \Phi_3 &= (\beta^* u_3 + \beta\, u_6) b_+ - (\beta u_2 + \beta^* u_5) b_-. \end{aligned} \tag{1.41}$$

Finally, we sequentially applied two unitary transformations to the Hamiltonian (1.35), one is with matrix T_1 and another one with matrix T_2. The total transformation matrix is the product of the two:

$$S = T_2T_1 = \begin{bmatrix} \alpha^* & 0 & 0 & \alpha & 0 & 0 \\ 0 & \beta b_+ & \beta^* b_- & 0 & \beta^* b_+ & \beta b_- \\ 0 & -\beta b_- & \beta^* b_+ & 0 & -\beta^* b_- & \beta b_+ \\ \alpha^* & 0 & 0 & -\alpha & 0 & 0 \\ 0 & \beta b_+ & -\beta^* b_- & 0 & -\beta^* b_+ & \beta b_- \\ 0 & -\beta b_- & -\beta^* b_+ & 0 & \beta^* b_- & \beta b_+ \end{bmatrix}, \quad SS^+ = 1, \tag{1.42}$$

Transformed Hamiltonian is block-diagonal:

$$\tilde{\tilde{H}} = SHS^+ = \begin{bmatrix} H^U & 0 \\ 0 & H^L \end{bmatrix}, \quad H^U = (H^L)^*. \tag{1.43}$$

The upper block of $\tilde{\tilde{H}}$ is given as

$$H^U = \begin{bmatrix} E_1 & K_t b_+ - iH_t b_- & -K_t b_- - iH_t b_+ \\ K_t b_+ + iH_t b_- & E_2 & -iH_t \\ iH_t b_+ - K_t b_- & iH_t & E_3 \end{bmatrix}, \tag{1.44}$$

where

$$E_1 = F, \; E_{2,3} = \frac{G+\lambda}{2} \pm \sqrt{\left(\frac{G+\lambda}{2}\right)^2 + 2\Delta_3^2} \tag{1.45}$$

The advantage of the representation (1.41) is that at the point $\mathbf{k} = 0$ the Hamiltonian (1.43) is diagonal, hence wave functions Φ_1, Φ_2, Φ_3 are band edge Bloch amplitudes of heavy hole (HH), light hole (LH), and crystal-field split (CH) bands, respectively. These basis functions have been used for the calculation of optical matrix elements and gain in GaN lasers [7].

1.4 Linear *k*-Terms in Wurtzite Nitrides

Matrix (1.33) is calculated using "quasi-cubic" basis functions (1.17). It contains linear-*k* matrix elements that stem from the first-order kp-term in the Hamiltonian (1.11). General symmetry considerations [1,8] allow additional linear-k terms in the Hamiltonian that are missed when we use the "quasi-cubic" basis. The more realistic basis accounts for mixing of $|Z>$ and $|iS>$ states as they belong to the same irreducible representation Γ_1 (see Fig. 1.3). Replacing basis functions $|Z>$ and $|iS>$ with their linear combinations

$$
\begin{aligned}
|Z'> &= \cos q\,|Z> + \sin q\,|iS> \\
|iS'> &= -\sin q\,|Z> + \cos q\,|iS>,
\end{aligned}
\tag{1.46}
$$

and recalculating matrix elements (1.18), we obtain additional linear-k terms. In a first-order approximation, these terms stem from two sources in the Hamiltonian (1.11): the non-relativistic kp-term and the relativistic spin–orbit contribution, H_2. First-order kp-interaction generates linear terms that appear in matrix elements (1.33): H_{24}, H_{34}, H_{68}, H_{78}:

$$
\begin{aligned}
&H_{24} = H_{78} = -H_{+}^{*},\ H_{34} = H_{68} = H_{-},\ H_{\pm} = \frac{\hbar^2}{2m_0} A_6 k_{+} k_z \pm A_7 k_{+}, \\
&A_7 = -\frac{i\hbar^2}{m_0\sqrt{2}} <X\left|\frac{\partial}{\partial x}\right|iS> \sin q,
\end{aligned}
\tag{1.47}
$$

Coefficient A_7 equals zero in the quasi-cubic approximation, $q = 0$. Linear-k terms proportional to coefficient A_7 lift the crossing of light-hole (E_2) and crystal field split-off (E_3) bands that occurs at the particular momentum along the c-direction [9]. The value of A_7 does not exceed 3×10^{-9} eV × cm across the (Ga–In–AI)N family [10]. Since A_7 term has non-relativistic origin, it does not cause the spin splitting in the electron spectrum and thus can be neglected as far as the spintronic properties of wurtzite materials are concerned.

Spin–orbit interaction H_2 generates linear-k matrix elements that mix initial basis spin states. These terms affect the spin structure of the energy spectrum and induce the zero magnetic field spin splitting in the bulk crystal, so they are relevant for spintronic properties of the material. It is instructive to consider additional matrix elements for the conduction band described by a rectangular submatrix in (1.33): H_{11}, H_{55}, H_{15}, and H_{51}. Matrix elements of type $<S\left|\frac{\partial V}{\partial x}\right|S>$ that follow from Eqs. (1.18) and (1.46), can be expressed as

$$<u_{n0}(r)\left|\frac{\partial V}{\partial x}\right|u_{n'0}(r)> = <u_{n0}(r)\left|\frac{\partial}{\partial x}H_0 - H_0\frac{\partial}{\partial x}\right|u_{n'0}(r)>$$
$$= (E_{n0} - E_{n'0})<u_{n0}(r)\left|\frac{\partial}{\partial x}\right|u_{n'0}(r)>, \quad (1.48)$$

and then matrix elements between the states of equal energy become zero:

$$<S\left|\frac{\partial V}{\partial x}\right|S> = <S\left|\frac{\partial V}{\partial y}\right|S> = <S\left|\frac{\partial V}{\partial z}\right|S> = 0$$
$$<Z\left|\frac{\partial V}{\partial x}\right|Z> = <Z\left|\frac{\partial V}{\partial y}\right|Z> = 0. \quad (1.49)$$

Non-zero terms give the conduction band matrix known as the Rashba Hamiltonian [11, 12]:

$$H_R = E_c + \frac{\hbar^2 k^2}{2m} + \alpha_R(k_y\sigma_x - k_x\sigma_y) = E_c + \frac{\hbar^2 k^2}{2m} + \sigma_R[\mathbf{k}\times\boldsymbol{\sigma}]_z,$$
$$\alpha_R = \frac{\hbar^2}{4c^2 m_0^2}<Z\left|\frac{\partial V}{\partial z}\right|Z>\sin^2 q. \quad (1.50)$$

Hamiltonian (1.50) describes the linear-k spin splitting caused by the bulk inversion asymmetry.

It should be noted that linear-k spin-split terms also exist in zinc blende materials for wave vectors in the [111] and [110] directions. However, the splitting appears only in the expanded set of initial basis functions that accounts for d-states [13]. In Chapter 2, the Hamiltonian (1.50) and the spin texture related to it will be discussed in more detail.

Problems

1.1 Show that the heavy hole band that follows from Hamiltonian (1.19) does not acquire a longitudinal effective mass.

1.2 The unitary transformation U acts on a wave function as $\Psi' = U\Psi'$. What does the transformed Hamiltonian look like?

1.3 Diagonalize matrix (1.39) and find coefficients (1.40).

1.4 Show that transformation matrix (1.39) is unitary.

References

1. Bir GL, Pikus GE (1974) *Symmetry and Strain Effects in Semiconductors*, Wiley, New York.
2. Anselm A (1981) *Introduction to Semiconductor Theory*, Prentice-Hall, New Jersey.
3. Chen GD, Smith M, Lin JY, Jiang HX, Wei SH, Asif Khan M, Sun CJ (1996) Fundamental optical transitions in GAN, *Appl Phys Lett*, **68**, 2784–2786.
4. Chuang SL, Chang CS (1996) $k \cdot p$ method for strained wurtzite semiconductors, *Phys Phys*, **B54**(4), 2491–2504.
5. Broido DA, Sham LJ (1985) Effective masses of holes at GaAs-AlGaAs heterojunctions, *Phys Rev*, **B31**, 888–892.
6. O'Reilly EP (1989) Valence band engineering in strained-layer structures, *Semicond Sci Technol*, **4**, 121–137.
7. Litvinov VI (2000) Optical transitions and gain in group-III nitride quantum wells, *J Appl Phys*, **88**, 5814–5820.
8. Lew Yan Voon LC, Willatzen M, Cardona M, Christensen NE (1996) Terms linear in k in the structure of wurtzite-type semiconductors, *Phys Rev*, **B53** (16), 10703–10714.
9. Kim K, Lambrecht WRL, Segall B, van Schilfgaarde M (1997) Effective masses and valence-band splitting in GAN and AlN, *Phys Rev*, **B56**(12), 7363–7375.
10. Dugdale DJ, Brand S, Abram RA (2000) Direct calculation of k-p parameters for wurtzite AlN, GAN, and InN, *Phys Rev*, **B61**(19), 12933–12938.
11. Rashba EI, Sheka VI (1959) Symmetry of energy bands in crystals of wurtzite-type II. Symmetry of the spin interactions, *Fizika Tverdogo Tela*, **1**(2), 162–176.
12. Casella RC (1960) Toroidal energy surfaces in crystals with wurtzite symmetry, *Phys Rev Lett*, **5**(8), 371–373.
13. Cardona M, Christensen NE, Fasol G (1988) Relativistic band structure and spin-orbit splitting of zinc-blende-type semiconductors, *Phys Rev*, **B38**, 3, 1810–1827.

Chapter 2

Rashba Hamiltonian

Two types of linear-k terms in the Hamiltonian were discussed in Chapter 1. One stems from the non-relativistic kp-interaction and thus plays no role in the spintronic properties of the material. The other one, given in Eq. (1.50), originates from spin-orbit interaction in bulk crystals and will be discussed here in more detail.

The coupling between electron spin and momentum affects the energy spectrum in two ways: It creates a spin-orbit split-off band twofold-degenerate in spin projection, as explained in Chapter 1 (Fig. 1.3), and it may generate additional linear and cubic momentum-dependent terms that lift spin degeneracy. Energy splitting that preserves spin degeneracy determines the rate of the Elliot–Yaffet spin relaxation that is the spin-flip-induced randomization of spin alignment in the course of momentum scattering. Linear and cubic spin splitting are at the origin of the Dyakonov–Perel spin relaxation mechanism [1] and also of electric dipole spin resonance (EDSR) that is induced by ac-electric field optical spin-flip transitions between two spin states [2, 3]. The spin-splitting terms in the Hamiltonian will be considered below.

Wide Bandgap Semiconductor Spintronics
Vladimir Litvinov

ISBN 978-981-4669-70-2 (Hardcover), 978-981-4669-71-9 (eBook)
www.panstanford.com

2.1 Bulk Inversion Asymmetry

Let us discuss the Rashba Hamiltonian obtained in Chapter 1. The Hamiltonian (1.50) is written in the spinor basis that is the pair of eigenfunctions of Pauli matrix $\sigma_z: |\uparrow> = \begin{pmatrix}1\\0\end{pmatrix}, |\downarrow> = \begin{pmatrix}0\\1\end{pmatrix}$. These spinors are not eigenfunctions of the Hamiltonian. That is why the Hamiltonian is non-diagonal in spin indexes.

So far we have been dealing with bulk wurtzite materials. The Hamiltonian

$$H_R = E_c + \frac{\hbar^2 k^2}{2m} + \alpha_R(k_y\sigma_x - k_x\sigma_y), \tag{2.1}$$

describes the bulk inversion asymmetry (BIA). Since Eq. (2.1) accounts for the conduction band only and neglects coupling to other bands, one can consider the model to be phenomenological. The free electron mass is replaced with an effective mass m to account for k^2-order contributions from the remote energy bands. Here, the Rashba coefficient α_R is the parameter that varies with structures and materials and can be calculated from a microscopic approach in each particular instance. Microscopic calculation of the Rashba coefficient in AlGaN quantum wells will be performed in Section 2.3 of this chapter and in Chapter 3.

Eigenvalues of H_R represent the spin-dependent electron energy spectrum:

$$E = E_c + \varepsilon(k), \varepsilon(k) = \frac{\hbar^2 k^2}{2m} \pm \alpha_R k, k = \sqrt{k_x^2 + k_y^2}, \tag{2.2}$$

Two branches of the spectrum correspond to the spin indexes as illustrated in Fig. 2.1:

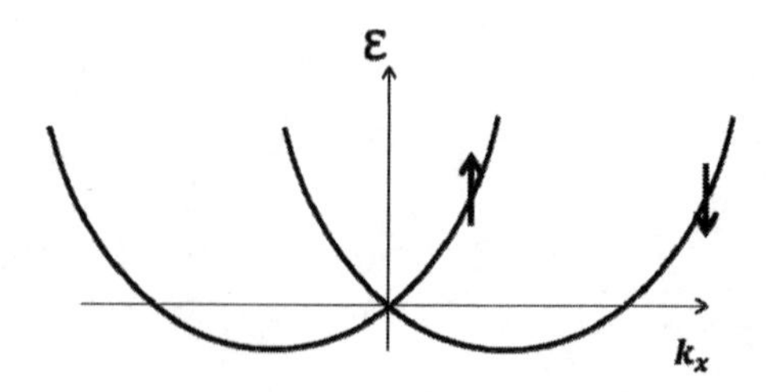

Figure 2.1 Rashba spin-splitting. Up- and down-arrows stand for the two spin projections.

Time reversal symmetry requires the twofold Kramers degeneracy to hold in the spin-split one-electron spectrum: $E\uparrow(\mathbf{k}) = E\downarrow(-\mathbf{k})$. Spin-up and down branches intersect at the Dirac point, $\varepsilon = 0$. Isoenergy surfaces are spherical at $\varepsilon > 0$ and toroidal at $\varepsilon < 0$.

Symmetry considerations require the spin-orbit part of the total bulk wurtzite Hamiltonian to be in the form [4, 5]:

$$H_{\mathrm{SO}} = \lambda[\boldsymbol{\sigma} \times \mathbf{k}]_z + \lambda_l[\boldsymbol{\sigma} \times \mathbf{k}]_z k_z^2 + \lambda_t[\boldsymbol{\sigma} \times \mathbf{k}]_z k^2, \tag{2.3}$$

where λ, λ_l, λ_t are phenomenological constants.

The effective Hamiltonian (2.3) contains a k-linear BIA Rashba term and also high-order contributions called, by analogy to zinc blende, k^3-Dresselhaus BIA terms. Spin splitting in a bulk wurtzite crystal that follows from (2.3) is absent for electrons moving in the c-direction. Parameters in Eq. (2.3) are the phenomenological constants to be determined by fitting theory to experimental data. In GaN they were found to be $\lambda < 4 \times 10^{-13}$ eV × cm [3], and $\lambda_l \approx 4\lambda_t$, $\lambda_t = 7.4 * 10^{-28}$ eV × cm^3 [6].

In GaAs-based zinc blende crystals, cubic basis functions (see Chapter 1) do not generate k-linear terms, so spin splitting starts from the Dresselhaus k^3-terms [7]:

$$\begin{aligned} H_{\mathrm{D}} &= \delta\boldsymbol{\sigma}\cdot\boldsymbol{\kappa}, \\ \kappa_x &= k_x(k_y^2 - k_z^2) \\ \kappa_y &= k_y(k_z^2 - k_x^2) \\ \kappa_z &= k_z(k_x^2 - k_y^2). \end{aligned} \tag{2.4}$$

Dresselhaus interaction Eq. (2.4) equals zero in the directions of all three principal cubic axes. BIA results in k-dependent zero magnetic field spin-splitting in bulk materials. No spin-splitting occurs in materials with inversion symmetry like Si-Ge. Besides, the BIA Rashba and Dresselhaus coefficients in wurtzite and cubic III-V materials are small and cannot be manipulated by an external force. The situation, however, looks different in low-dimensional structures like heterostructures, quantum wells, and quantum dots as the engineered structure inversion asymmetry (SIA) may play the role of BIA in bulk material. So, low-dimensional structures may generate engineered electron spin-splitting that depends on the geometrical parameters of the structure. Linear *k*-spin splitting in low-dimensional structures will be discussed below.

2.2 Structure Inversion Asymmetry

In artificial structures, for instance, quantum wells (QW), the asymmetric confining field creates SIA that can be manipulated by an applied voltage as well as by geometrical parameters of the structure. Two-dimensional electrons under SIA conditions reveal linear spin-splitting and can be described by a Rashba Hamiltonian that formally coincides with that in the bulk Eq. (2.1), where the *z*-axis is perpendicular to the QW plane rather than the hexagonal axis in the bulk [8]. It is instructive to compare the spin configurations induced by spin-orbit interaction in zinc blende and wurtzite quantum wells.

Zinc blende QW. The linear spin-orbit interaction in zinc blende quantum well (ZB-QW) grown in the [001] direction comprises two contributions: the Rashba term (2.1) and the Dresselhaus term (2.4) modified by one-dimensional electron confinement:

$$H_{\mathrm{ZB}} = \alpha_R(k_y\sigma_x - k_x\sigma_y) + \alpha_D(k_x\alpha_x - k_y\alpha_y),\ \alpha_D = -\delta < k_z^2 >, \tag{2.5}$$

where α_D and α_R are the Dresselhaus and Rashba coefficients, respectively. It is convenient to represent H_{ZB} in the form

$$H_{\mathrm{ZB}} = \frac{1}{\sqrt{2}}\begin{pmatrix} 0 & \gamma k e^{i\theta} \\ \gamma k e^{-i\theta} & 0 \end{pmatrix}, \tag{2.6}$$

where

$$\gamma = \sqrt{\alpha_R^2 + \alpha_D^2 + 2\alpha_R\alpha_D \sin 2\varphi},$$
$$\tan\theta = \frac{\alpha_R \cos\varphi + \alpha_D \sin\varphi}{\alpha_R \sin\varphi + \alpha_D \cos\varphi},\ \tan\varphi = k_y/k_x. \tag{2.7}$$

In order to find eigenvalues and eigenfunctions, one has to diagonalize the Hamiltonian by the unitary transformation. The unitary transformation that diagonalizes H_{ZB} up to an overall phase factor may be written as

$$U_{\mathrm{ZB}} = \frac{1}{\sqrt{2}} \begin{pmatrix} -e^{i\theta} & e^{i\theta} \\ 1 & 1 \end{pmatrix},$$

$$|\mathrm{Det}[U_{\mathrm{ZB}}]| = 1,\; U_{\mathrm{ZB}}U_{\mathrm{ZB}}^{+} = 1. \tag{2.8}$$

The orthonormal eigenspinors and diagonalized Hamiltonian can be calculated as

$$u_{\pm} = U_{\mathrm{ZB}}|\uparrow\downarrow> = \frac{1}{\sqrt{2}} \begin{pmatrix} \mp e^{i\theta} \\ 1 \end{pmatrix},$$

$$\widetilde{H} = U_{\mathrm{ZB}}^{+} H_{\mathrm{ZB}} U_{\mathrm{ZB}} = \begin{pmatrix} -\gamma k & 0 \\ 0 & \gamma k \end{pmatrix}. \tag{2.9}$$

The spin splitting resulting from the Hamiltonian (2.9) is given as $\Delta\varepsilon = 2\gamma k$.

The components of the average spin vectors $S_{\pm}$ in $u_{\pm}$ states can be found by calculating the matrix elements $\boldsymbol{S}_{\pm} = -\boldsymbol{S}_{\mp} = \frac{1}{2} <u_{\pm}|\boldsymbol{\sigma}|u_{+}>$:

$$S_{x\pm} = \mp\frac{1}{2}\cos\theta = \mp\frac{\alpha_R \sin\varphi + \alpha_D \cos\varphi}{2\sqrt{\alpha_R^2 + \alpha_D^2 + 2\alpha_R\alpha_D \sin 2\varphi}},$$

$$S_{y\pm} = \pm\frac{1}{2}\sin\theta = \pm\frac{\alpha_R \cos\varphi + \alpha_D \sin\varphi}{2\sqrt{\alpha_R^2 + \alpha_D^2 + 2\alpha_R\alpha_D \sin 2\varphi}}. \tag{2.10}$$

The average spin direction is linked to instantaneous electron momentum and defines the direction of the effective magnetic field $\mathbf{B}_{\pm}(\mathbf{k})$ acting on an electron. The directions of the magnetic field are opposite for electrons with opposite wave vectors or spins (±): $\mathbf{B}_{\pm}(-\mathbf{k}) = -\mathbf{B}_{\pm}(\mathbf{k})$, $\mathbf{B}_{\pm}(\mathbf{k}) = -\mathbf{B}_{\mp}(\mathbf{k})$. The orientation of the electron spin (or effective magnetic field), depending on its momentum, is illustrated in Figs. 2.2 and 2.3.

Conditions $\alpha_D = \pm\alpha_R$ make spin orientation independent of momentum in a wide area of the phase space. This follows from Eq. (2.7) and is illustrated in Fig. 2.3 by the parallel spin vectors independent of phase φ. So, the change of the momentum does not rotate the spin or, in other words, the spin-orbit interaction does not preclude spin conservation against all forms of spin-independent scattering [9].

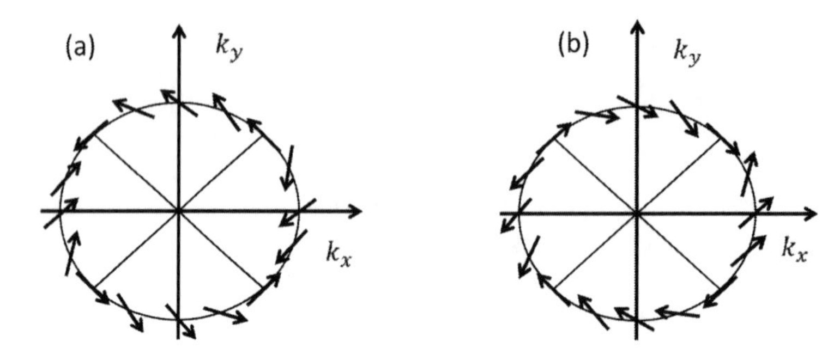

Figure 2.2 Electron spin configuration in spin states (+) (a) and (–)(b), $\alpha_D \neq \alpha_R$.

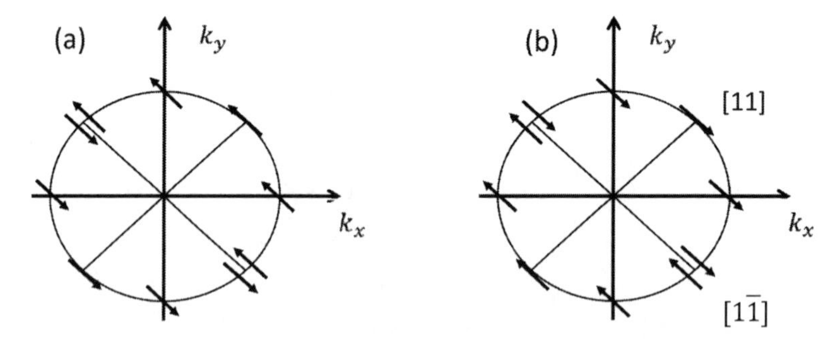

Figure 2.3 Electron spin configuration in spin states (+)(a) and (–)(b), $\alpha_D = \alpha_R$.

The temporal evolution of the spin texture in the case $\alpha_D = \pm\alpha_R$ can be reconstructed by inspecting the details in Fig. 2.3. Since the effective magnetic field is perpendicular to the [11] momentum and lies in the (x, y) plane, the z-component of the electron spin rotates in the plane normal to (x, y) while propagating along the [11] direction. The $[1\bar{1}]$ momentum direction is special as the effective magnetic field changes sign and equals zero for electrons moving in this direction. So, no spin precession occurs in the $[1\bar{1}]$ direction and the electronic spin is conserved. This special case is termed 'persistent spin helix' [10].

When $\alpha_R \to 0$ the Dresselhaus spin texture can be found by substituting $\cos\theta \to \cos\varphi$, $\sin\theta \to \sin\varphi$. In this limit, the transformation matrix, eigenspinors and the transformed Hamiltonian follow:

$$U_{\text{ZB}} = \frac{1}{\sqrt{2}}\begin{pmatrix} -e^{i\varphi} & e^{i\varphi} \\ 1 & 1 \end{pmatrix},$$

$$u_{\pm} = U_{\text{ZB}}\ |\uparrow\downarrow> = \frac{1}{\sqrt{2}}\begin{pmatrix} \mp e^{i\varphi} \\ 1 \end{pmatrix},$$

$$\widetilde{H} = U_{\text{ZB}}^{+} H_{\text{ZB}} U_{\text{ZB}} = \begin{pmatrix} -\alpha_{\text{D}} k & 0 \\ 0 & \alpha_{\text{D}} k \end{pmatrix} \tag{2.11}$$

In the $u_{\pm}$ states, the average spin components are given as

$$S_{x\pm} = \mp\frac{1}{2}\cos\varphi;\ S_{y\pm} = \pm\frac{1}{2}\sin\varphi, \tag{2.12}$$

and shown in Fig. 2.4.

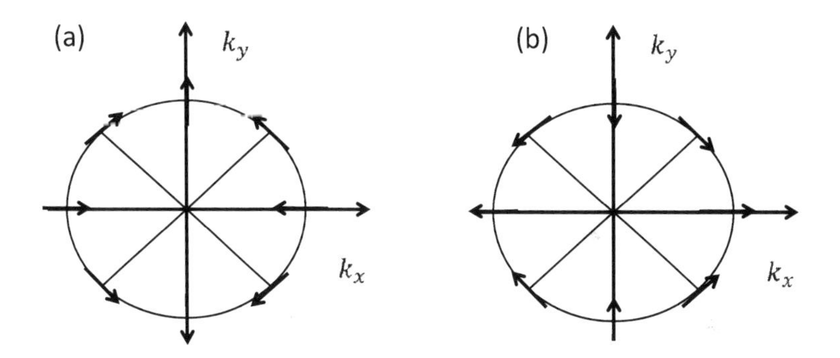

Figure 2.4 Dresselhaus spin configurations in spin states (+)(a) and (−)(b).

Wurtzite QW. In a wurtzite quantum well (W-QW) with the electron confinement in the c-direction, the effective linear spin-orbit coupling follows from Eq. (2.3). Under confinement in the z-direction, k_z can be replaced with its quantized values k_n, n = 1, 2, ... For instance, in the ground state the replacement can be made as

$$k_z \rightarrow <k_z> = -i<\nabla_z> \equiv -i\int \Phi^{*}\nabla_z \Phi dz = 0,\ k_z^2 \ \rightarrow\ <k_1^2> = (\pi/L)^2, \tag{2.13}$$

where $\Phi(z)$ is the ground state wave function in a QW of width L and infinite barrier height. After that, Eq. (2.3) can be written as

$$H_{\mathrm{W}} = \alpha_{\mathrm{eff}}\,(\alpha_x k_y - k_x\sigma_y),$$
$$\alpha_{\mathrm{eff}} = \alpha_R + \alpha_{\mathrm{BIA}},$$
$$\alpha_{\mathrm{BIA}} = \lambda + \lambda_l < k_z^2 > . \tag{2.14}$$

The Hamiltonian (2.14) is a limit of (2.5) when $\alpha_D \to 0$. The limit implies following relation between θ and the electron phase $\cos\theta \to \sin\varphi$, $\sin\theta \to \cos\varphi$, so the transformation matrix, eigenspinors, and transformed Hamiltonian can be written as

$$U_{\mathrm{W}} = \frac{1}{\sqrt{2}}\begin{pmatrix} -ie^{-i\varphi/2} & -ie^{-i\varphi/2} \\ 1 & 1 \end{pmatrix},\ u_{\pm} = U_{\mathrm{W}}|\uparrow\downarrow> = \frac{1}{\sqrt{2}}\begin{pmatrix} \mp ie^{-i\varphi} \\ 1 \end{pmatrix},$$
$$\widetilde{H} = U_{\mathrm{W}}^{+} H_{\mathrm{W}} U_{\mathrm{W}} = \begin{pmatrix} -\alpha_{\mathrm{eff}}k & 0 \\ 0 & \alpha_{\mathrm{eff}}k \end{pmatrix} \tag{2.15}$$

Then the spin splitting is equal to $\Delta\varepsilon = 2\alpha_{\mathrm{eff}}k$. In the $u_{\pm}$ states, the spin components are given as

$$S_{x\pm} = \mp\frac{1}{2}\sin\varphi;\ S_{y\pm} = \pm\frac{1}{2}\cos\varphi, \tag{2.16}$$

and their directions are shown in Fig. 2.5.

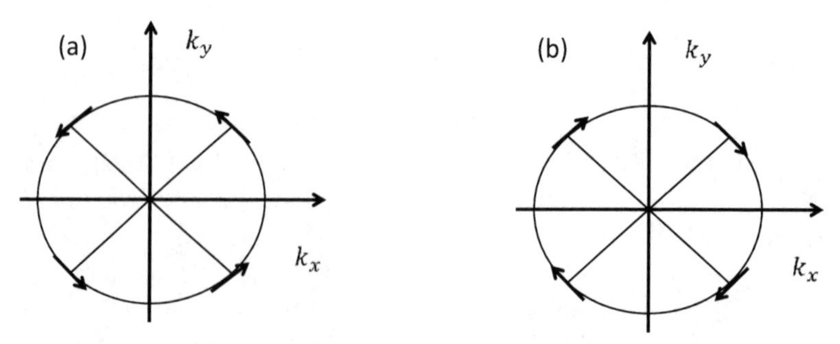

Figure 2.5 Rashba spin configurations in spin states (+) (a) and (–) (b).

The two-dimensional (±) electron states described by the Rashba–Dresselhaus Hamiltonians are often referred in the literature as states with different helicity. Also, in topological insulators (see Chapters 7 and 8) electron spin states locked to the direction of propagation are referred as helical states. In Rashba and Dresselhaus electron gas this usage is often confusing, as the helicity in quantum field theory has the exact meaning of an eigenvalue of the operator $\sigma\boldsymbol{k}/k$ that is a projection of the average spin on the direction of particle momentum. It is true that sometimes the finite helicity can be associated with (±) states when spin projection on the direction of momentum is finite. However, inspecting Figs. 2.2–2.5, we can conclude that in some cases, for example, in Rashba electron gas (see Fig. 2.5), the helicity equals zero and cannot be a quantum number that characterizes (±) electron states. The same can be said about Rashba– Dresselhaus gas where the helicity becomes zero in certain momentum directions. It is safe to call (±) spin states bearing in mind that they are linear combinations of the initial spin states of an electron before any spin-dependent interaction has been taken into account.

2.3 Microscopic Theory of Rashba Spin Splitting in GaN

Typical values of the Rashba coefficient α_R in InAs/GaSb, InP/InGaAs and InAs/InGaAs QWs are (0.6–4) × 10^{-11} eVm [11, 12], which gives zero magnetic field electron spin-splitting energy on the order of (1–10) meV depending on the doping level. In order to engineer the spin splitting by manipulation of QW geometrical parameters, the well/barrier lattice mismatch strain, and external electric field, it is important to know the relation of the Rashba coefficient to the initial microscopic parameters of the material. So far in this chapter the Rashba Hamiltonian (2.1) has been derived taking into account the conduction band only and neglecting the influence of other bands. As a result, the Rashba coefficient and the effective mass appear as phenomenological constants. Thus the relation of these constants to bulk material parameters is missed. In this section we present the microscopic derivation of linear spin-splitting that relates the Rashba coefficient in wurtzite GaN/AlGaN heterostructures to parameters of the bulk crystal.

Let us consider a heterostructure grown in the z-direction as a stressed (Al,In,Ga)N layer on a thick and relaxed GaN substrate, or QW of either AlGaN/GaN/AlGaN or GaN/InGaN/GaN-type. The lattice constant mismatch introduces the strain, which enters the

Hamiltonian (1.19), renormalizing the diagonal matrix elements as follows:

$$H = \begin{bmatrix} E_c & -\frac{P_2k_+}{\sqrt{2}} & \frac{P_2k_-}{\sqrt{2}} & P_1k_z & 0 & 0 & 0 & 0 \\ -\frac{P_2k_-}{\sqrt{2}} & F & 0 & 0 & 0 & 0 & 0 & 0 \\ \frac{P_2k_+}{\sqrt{2}} & 0 & G & 0 & 0 & 0 & 0 & \sqrt{2}\Delta_3 \\ P_1k_z & 0 & 0 & \lambda & 0 & 0 & \sqrt{2}\Delta_3 & 0 \\ 0 & 0 & 0 & 0 & E_c & \frac{P_2k_-}{\sqrt{2}} & -\frac{P_2k_+}{\sqrt{2}} & P_1k_z \\ 0 & 0 & 0 & 0 & \frac{P_2k_+}{\sqrt{2}} & F & 0 & 0 \\ 0 & 0 & 0 & \sqrt{2}\Delta_3 & -\frac{P_2k_-}{\sqrt{2}} & 0 & G & 0 \\ 0 & 0 & \sqrt{2}\Delta_3 & 0 & P_1k_z & 0 & 0 & \lambda \end{bmatrix}$$

$$F = E_{v0} + \Delta_1 + \Delta_2 + S_1 + S_2 + V(z),\ G = \Delta_1 - \Delta_2 + S_1 + S_2 + V(z),$$
$$S_1 = D_1\varepsilon_{zz} + D_2(\varepsilon_{xx} + \varepsilon_{yy});\ S_2 = D_3\varepsilon_{zz} + D_4(\varepsilon_{xx} + \varepsilon_{yy}),$$
$$\lambda = S_1 + V(z), \tag{2.17}$$

where D_i are the deformation potentials, $\varepsilon_{xx} = \varepsilon_{yy} = (a_0 - a)/a$, $\varepsilon_{zz} = -(2C_{13}/C_{33})\varepsilon_{xx}$ are the strain components, C_{13} and C_{33} are the elastic coefficients, a_0 and a are the in-plane lattice constants of the substrate and layer, respectively. We assume that z-dependent parameters E_c and E_{v0} account for the heterostructure band offsets. Here $V(z)$ includes contributions from an external electric field (bias) and a self-consistent potential in an inhomogeneous structure. Shear strain components are neglected.

The matrix Schrödinger equation for the 8-component envelope wave function has the form $H\varphi = \varepsilon\varphi$, where

$$\varphi = (\varphi_1, \varphi_2, \varphi_3, \varphi_4, \varphi_5, \varphi_6, \varphi_7, \varphi_8)^T, \tag{2.18}$$

where $(...)^T$ means transpose the row into a column. Components $\varphi_{1,2}$ correspond to conduction band amplitudes with spin $\uparrow\downarrow$, respectively.

In a heterostructure grown along the c-axis, one has to replace k_z with $-i\partial/\partial z$ and keep band parameters z-dependent. The system of eight equations can be exactly decoupled into a two-component conduction envelope function $\begin{pmatrix}\varphi_1\\ \varphi_2\end{pmatrix}$ as

$$\begin{bmatrix} E_c + V(z) + \dfrac{\hbar^2 k^2}{2m} - \dfrac{\hbar^2}{2}\dfrac{\partial}{\partial z}\dfrac{1}{m_z}\dfrac{\partial}{\partial z} - \varepsilon & iP_1P_2k_-\dfrac{\partial\beta}{\partial z} \\ -iP_1P_2k_+\dfrac{\partial\beta}{\partial z} & E_c + V(z) + \dfrac{\hbar^2 k^2}{2m} - \dfrac{\hbar^2}{2}\dfrac{\partial}{\partial z}\dfrac{1}{m_z}\dfrac{\partial}{\partial z} - \varepsilon \end{bmatrix}\begin{pmatrix}\varphi_1\\ \varphi_2\end{pmatrix} = 0 \tag{2.19}$$

$$m^{-1} = \frac{1}{m_0} + \frac{2P_2^2}{\hbar^2}\frac{2\Delta_3^2 - (\lambda-\varepsilon)(F+G-2\varepsilon)}{(F-\varepsilon)[(G-\varepsilon)(\lambda-\varepsilon) - 2\Delta_3^2]},$$

$$m_z^{-1} = \frac{1}{m_0} + \frac{2P_1^2}{\hbar^2}\frac{(\varepsilon - G)}{(G-\varepsilon)(\lambda-\varepsilon) - 2\Delta_3^2},$$

$$\beta = \frac{\Delta_3}{(G-\varepsilon)(\lambda-\varepsilon) - 2\Delta_3^2} \tag{2.20}$$

In a relaxed layer, the strain is zero and the electron effective masses in Eq. (2.20) coincide with those obtained in Ref. [13], assuming that the reference energy $E_{v0} = 0$, $\varepsilon \to E_c = E_g + \Delta_1 + \Delta_2$, where E_g is the bandgap.

The diagonal part of the Hamiltonian (2.19) has spin degenerate eigenfunctions $\Phi_1 = \Phi_2 = \Phi$. Non-diagonal terms in Eq. (2.19) lift the spin degeneracy. The energy difference between spin-up and spin-down conduction states (spin-splitting) can be calculated as

$$\Delta\varepsilon = 2\alpha_R k,$$

$$\alpha_R = P_1P_2\left\langle\frac{\partial\beta}{\partial z}\right\rangle,$$

$$\langle f\rangle = \int \Phi^*(z)f(z)\Phi(z)dz \tag{2.21}$$

Let us analyze the general expression for the Rashba parameter α_R in a generic AlGaN/GaN/AlGaN QW of width L. Coefficient β depends on z and can be written as the sum of three terms each corresponding to a region i representing the left barrier L, the well W, and the right barrier R as follows:

$$\beta(z) = \beta_L[1 - \theta(z)] + \beta_W[\theta(z) - \theta(z - L)] + \beta_R\theta(z - L), \tag{2.22}$$

where $\theta(z)$ is the step function. The matrix element with a QW ground state wave function determines the Rashba coefficient and is expressed as

$$\left\langle \frac{\partial \beta}{\partial z} \right\rangle = \Phi^2(0)(\beta_W - \beta_L) - \Phi^2(L)(\beta_W - \beta_R) + (B_L F_L) + \langle B_R F_R \rangle + \langle B_W F_W \rangle, \tag{2.23}$$

where

$$\beta_i = \frac{\Delta_3}{(E_g + 2\Delta_2 - S_1 - S_2 + \varepsilon - V_i)(E_g + \Delta_1 + \Delta_2 - S_1 + \varepsilon - V_i) - 2\Delta_3^2},$$

$$B_i = \frac{\Delta_3[2E_g + \Delta_1 + 3\Delta_2 - 2S_1 - S_2 + 2(\varepsilon - V_i)]}{\{(E_g + 2\Delta_2 - S_1 - S_2 + \varepsilon - V_i)(E_g + \Delta_1 + \Delta_2 - S_1 + \varepsilon - V_i - 2\Delta_3^2\}^2},$$

$$F_i = \frac{\partial V_i}{\partial z}. \tag{2.24}$$

The reference energy in Eqs. (2.23) and (2.24) is the bottom of the conduction band. Each average value contains integration over the corresponding region i excluding interfaces located at $z = 0$ and $z = L$. Potential jumps and offsets of band parameters at the interfaces contribute to the first two terms of Eq. (2.23). The energy parameter ε in all B_i and β_i should be taken at the edge of the ground electron level in the well, $k = 0, \varepsilon \to \varepsilon_1$. All other parameters in Eqs. (2.23) and (2.24) take the values attributed to the corresponding layer i.

If the barrier height tends to infinity, the coupling coefficient takes the form

$$\left\langle \frac{\partial \beta}{\partial z} \right\rangle = \langle B_W F_W \rangle \tag{2.25}$$

The Rashba coefficient $\alpha_R = P_1 P_2 \left\langle \frac{\partial \beta}{\partial z} \right\rangle$ is proportional to the average electric field in the well if B_W does not depends on z.

Otherwise the integral in (2.25) includes the product of the electric field and the combination of z-dependent material parameters. As follows from (2.23), no spin-splitting occurs in a symmetric QW where the following conditions hold, $\Phi^2(0)(\beta_W - \beta_L) = \Phi^2(L)(\beta_W - \beta_R)$ and $B_L F_L + B_R F_R = B_W F_W = 0$.

Comparing the BIA and SIA in zinc-blende (ZB) and wurtzite (W) materials, it is important to realize the essential difference between the two. In a symmetric and unbiased ZB [100]-oriented QW the SIA does not exist and the Rashba coupling becomes zero. On the other hand, even a structurally symmetric and unbiased W-QW is electrically asymmetric due to built-in polarization fields. This results in a non-zero Rashba coupling coefficient, which depends on internal fields. Another difference is that the c-axis oriented two-dimensional electrons in a W-structure feel the spin-orbit interaction in a different way compared to a ZB structure. It is known that $\boldsymbol{k}$-dependent electron spin configuration in ZB [100]-QW depends on the relation between the competing Dresselhaus and Rashba terms. In a c-axis oriented W-structure both the Rashba and the Dresselhaus-type terms induce the same electron spin configuration.

The Rashba parameter, calculated with (2.21) and (2.23), contains all of the information on the material and geometrical parameters of the structure that allow engineering the zero magnetic field spin-splitting in quasi-2D structures for various spintronic applications. In Chapter 3 a remotely-doped heterostructure and a polarization-distorted QW will be discussed in more detail.

Problems

2.1 The spinor corresponding to one of two spin projections on axis $\boldsymbol{n}$ pointing in the (θ, φ) – direction has the form:

$$|\theta, \varphi\rangle = \begin{pmatrix} \cos(\theta/2) \\ \sin(\theta/2)\exp(-i\varphi) \end{pmatrix}.$$

Find the spinor after a 2π θ-rotation.

2.2 Calculate the spectrum and average spin components in a two-dimensional Rashba electron gas in a perpendicular Zeeman magnetic field.

References

1. Pikus GE, Titkov AN (1984) Spin relaxation under optical orientation in semiconductors, in *Optical Orientation*, (Meier F, Zakharchenya BP, ed), North Holland, Amsterdam, pp. 73–131.
2. Rashba EI, Sheka VI (1991) Electric-dipole spin resonances, in: *Landau Level Spectroscopy*, North Holland, Amsterdam, pp. 131–206.
3. Wołoś A, Wilamowski Z, Skierbiszewski C, Drabinska A, Lucznik B, Grzegory I, Porowski S (2011) Electron spin resonance and Rashba field in GAN-based materials, *Physica B, Condensed Matter*, **406**(13), 2548–2554.
4. Bir GL, Pikus GE (1974) *Symmetry and Strain Effects in Semiconductors*, Wiley, New York.
5. Zorkani I, Kartheuser E, Resonant magnetooptical spin transitions in zinc-blende and wurtzite semiconductors, *Phys Rev*, **B53**, 4, 1871–1880.
6. Wang WT, Wu CL, Tsay SF, Gau MH, Lo I, Kao HF, Jang DJ, Chiang JC, Lee ME, Chang YC, Chen CN, Hsueh HC (2007) Dresselhaus effect in bulk wurtzite materials, *Appl Phys Lett*, **91**, 082110.
7. Dresselhaus G (1955) Spin-orbit coupling effects in zinc blende structures, *Phys Rev*, **100**, 580–586.
8. Rashba EI (1960) Properties of semiconductors with an extremum loop. 1. Cyclotron and combinational resonance in a magnetic field perpendicular to plane of the loop. *Sov Phys Solid State*, **2**, 1109–1122; Bychkov YA, Rashba EI (1984) Oscillatory effects and the magnetic susceptibility of carriers in inversion layers, *J. Phys. C: Solid State Phys.*, **17**, 6039–6045; Bychkov YA, Rashba EI (1984) Properties of a 2D electron gas with lifted spectral degeneracy, *JETP Lett*, **39**(2), 78–81.
9. Schliemann J, Carlos Egues J, Loss D (2003) Nonballistic spin-field-effect transistor, *Phys Rev Lett*, **90**(14), 146801.
10. Koralek JD, Weber CP, Orenstein J, Bernevig BA, Zhang S-C, Mack S, Awschalom DD (2009) Emergence of the persistent spin helix in semiconductor quantum wells, *Nature*, **458**, 610–614.
11. De Andrada e Silva EA, La Rocca GC, Bassani F (1994) Spin-split subbands and magneto-oscillations in III-V asymmetric heterostructures, *Phys Rev*, **B50**, 8523–8533.

12. Awshalom DD, Loss D, Samarth N (eds) (2002) *Semiconductor Spintronics and Quantum Computation*, Springer, Berlin, Germany.

13. Chuang SL, Chang CS (1996) $k \cdot p$ method for strained wurtzite semiconductors, *Phys Rev*, **B54**, 2491–2504.

Chapter 3

Rashba Spin Splitting in III-Nitride Heterostructures and Quantum Wells

Gate-voltage manipulation of electron spins in semiconductor heterostructures is the subject of intensive study with regard to semiconductor spintronic devices such as spin transistors, spin light-emitting diodes, and quantum computers [1, 2].

The general expression for the structure inversion asymmetry (SIA) Rashba coupling parameter in GaN/AlGaN heterostructures was obtained in Chapter 2. Basically, the spin-orbit coupling parameter α_R decreases as the bandgap increases. Thus, the spin splitting in narrow bandgap semiconductors like InGaAs is expected to be larger than that in larger bandgap materials. However, in a wide bandgap III-nitride heterostructure, the spin splitting, since it is approximately proportional to an average electric field in the growth direction, is affected by the in-built electric field. It is known that lattice polarization strongly affects the performance of GaN-based electronic and optoelectronic devices [3, 4]. In addition, the spin splitting depends not only on the Rashba parameter α_R but also on the Fermi level in degenerate conduction or the valence band. That is why strong polarization doping effect in III-nitrides, that is the increase in carrier density in the channel due to internal electric fields, may enhance overall spin splitting making it comparable to that found in narrow-gap III-V structures. One more

Wide Bandgap Semiconductor Spintronics
Vladimir Litvinov

ISBN 978-981-4669-70-2 (Hardcover), 978-981-4669-71-9 (eBook)
www.panstanford.com

reason to take III-nitrides into consideration is the experimental data on the narrow bandgap in InN (0.69 eV) [5] that might include InGaN alloys in the spintronic material family. Besides, InGaN quantum well can be grown as topological insulator (see Chapter 7) that is the class of materials which reveals strong coupling between spin and transport characteristics [6].

In this chapter we discuss the spintronic capabilities of wurtzite III-nitride heterostructures and quantum wells, namely, the Rashba spin-orbit coupling parameter and conduction-band spin splitting, determine their sensitivity to gate voltage, and establish the relation between the polarization field and electron spin splitting. Since GaN–InN–AIN is the pyroelectric material family, spontaneous and piezoelectrical polarizations may affect the electrically driven magnetism of confined electrons. In Section 3.1 we relate the internal built-in electric field to the interface electrical polarization caused by spontaneous and lattice-mismatch-induced piezoelectric contributions. Section 3.2 deals with the high-electron-mobility transistor (HEMT) structure, where polarization-induced doping results in a high density of two-dimensional (2D) electrons. Rashba interaction in polarization-doped heterostructure is explored in Section 3.3. Section 3.4 deals with structurally symmetric quantum wells whose conduction band profile is distorted by polarization fields.

3.1 Spontaneous and Piezoelectric Polarization

Bulk wurtzite crystals belong to the pyroelectric symmetry class 6 mm. This means that an external electric field induces an internal stress proportional to the electric field (piezoelectric effect). The converse effect is that a deformation creates an internal electric field, proportional to the deformation. In pyroelectric crystals spontaneous polarization (dipole moment per unit volume) exists at any temperature until the melting point. This differentiates this material from ferroelectric material where the spontaneous polarization occurs as a phase transition when the temperature decreases below its critical value.

In devices made of III-nitrides the stress originates from a lattice mismatch at an interface and also from processing related phenomena. In a crystal of wurtzite symmetry, three independent components of piezoelectric tensor d determine the relation between the components of the stress tensor σ and the polarization vector **P** [7]:

$$P_x = \frac{1}{2} d_{15}\sigma_{zx};\ P_y = \frac{1}{2} d_{15}\sigma_{zy};\ P_z = d_{31}(\sigma_{xx} + \sigma_{yy}) + d_{33}\sigma_{zz} \tag{3.1}$$

If a heterostructure comprises a substrate and an epitaxial layer grown in the [0001] direction, the lattice-mismatch stress has components in the hexagonal plane only. Thus d_{31} is the only relevant piezoelectric constant that relates the in-plane stress to c-axis oriented polarization:

$$P_{\text{pz}} = d_{31}(\sigma_{xx} + \sigma_{yy}). \tag{3.2}$$

It is more convenient to express the polarization in terms of components of the deformation tensor ϵ since the deformation is directly determined by the lattice mismatch between the substrate and the pseudomorphic epitaxial layer that accommodates the full lattice-mismatch stress. The stress-deformation relation is given by Hook's Law:

$$\sigma_i = \sum_j C_{ij}\varepsilon_j, i, j = 1, 2, \ldots 6, \tag{3.3}$$

where C_{ij} is the 6 × 6 matrix of elastic constants. The components of stress and deformation tensors are written in Voigt notations: $xx \to 1$, $yy \to 2$, $zz \to 3$, $yz, zy \to 4$, $xz, zx \to 5$, $xy, yx \to 6$.

Wurtzite symmetry dictates that elastic constants be given as

$$C_{ij} = \begin{pmatrix} C_{11} & C_{12} & C_{13} & 0 & 0 & 0 \\ C_{12} & C_{11} & C_{13} & 0 & 0 & 0 \\ C_{13} & C_{13} & C_{33} & 0 & 0 & 0 \\ 0 & 0 & 0 & C_{44} & 0 & 0 \\ 0 & 0 & 0 & 0 & C_{44} & 0 \\ 0 & 0 & 0 & 0 & 0 & \frac{1}{2}(C_{11} - C_{12}) \end{pmatrix} \tag{3.4}$$

The epitaxial layer grown in the [0001]-direction, remains stress-free in that direction: $\sigma_3 = 0$. Expressed in terms of deformation with Eq. (3.3), the stress-free condition transforms into $C_{31}\varepsilon_1 + C_{32}\varepsilon_2 + C_{33}\varepsilon_3 = 0$. Then with Eq. (3.4) it becomes $C_{13}(\varepsilon_1 + \varepsilon_2) + C_{33}\varepsilon_3 = 0$. In-plane biaxial strain implies that $\varepsilon_1 = \epsilon_2$, and finally, the relation between in-plane and vertical strain components is

$$\varepsilon_3 = -2\frac{C_{13}}{C_{33}}\varepsilon_1, \tag{3.5}$$

where $2\frac{C_{13}}{C_{33}}$ is Poisson's ratio. Strain ε_1 can be written in terms of the mismatch between in-plane lattice constants of substrate a and layer a_0 materials: $\varepsilon_1 = \frac{a - a_0}{a_0}$. Strain $\varepsilon_3 = \frac{c - c_0}{c_0}$ is the mismatch between c-direction lattice constant in a stressed layer c and the fully relaxed layer c_0.

The definition of strain implies that the tensile strain is positive. The electrical polarization calculated with Eqs. (3.2), (3.3), and (3.5), is given as

$$P_{\mathrm{pz}} = 2d_{31}\varepsilon_1\left(C_{11} + C_{12} - 2\frac{C_{13}^2}{C_{33}}\right) = 2\varepsilon_1\left(e_{31} - e_{33}\frac{C_{13}}{C_{33}}\right), \tag{3.6}$$

where e_{31} and e_{33} are the piezoelectric constants that relate the polarization to a strain. The sign of P_z is referenced to the positive [0001] direction: from cation (Ga, Al, In) to the nearest neighbor anion (N) along the c-axis (see Chapter 1, Fig. 3.1). Since the product $d_{31}\left(C_{11} + C_{12} - 2\frac{C_{13}^2}{C_{33}}\right)$ and the bracket $\left(e_{31} - e_{33}\frac{C_{13}}{C_{33}}\right)$ are both negative in the (Ga, Al, In)N family, the sign of the polarization is determined by the sign of the strain: positive if $\varepsilon_1 < 0$ (compressive strain) and negative if $\varepsilon_1 > 0$ (tensile strain).

Total polarization consists of piezo-(P_{pz}) and spontaneous (P_{sp}) components. Spontaneous polarization is not related to stress and exists even in a fully relaxed layer. The direction of the spontaneous dipole moment in a unit cell always points from the anion to the nearest cation along the c-axis that is opposite to the [0001] direction, so the spontaneous dipole moment is always negative. As for the direction of the spontaneous polarization with respect to a substrate, it depends on what type of layer is under consideration. If an epitaxial layer terminates at the Ga plane

(Ga-face), the [0001] direction points toward the free surface while the dipole moment has the opposite direction toward the substrate. Layers of opposite polarity terminate at the N-face, the [0001] direction points toward the substrate, and the spontaneous polarization being negative is directed toward the free surface.

3.2 Remote and Polarization Doping

Polarization introduces built-in electric fields in the heterostructure and thus affects electric, magnetic, and optical properties of various GaN-based semiconductor devices. In order to estimate the electric fields and the polarization induced charge accumulating at the interface, we consider a typical example of modulation doped $Al_xGa_{1-x}N/GaN$ heterostructure shown in Fig. 3.1; x is the alloy composition, ΔE_C is the conduction band offset, φ_B is the Schottky barrier height, E_F is the Fermi energy, V_g is the gate voltage, and q is the elementary charge. Ionized donors in the barrier are shown by + signs placed close to the AlGaN conduction band edge, ε_d is the donor bound energy.

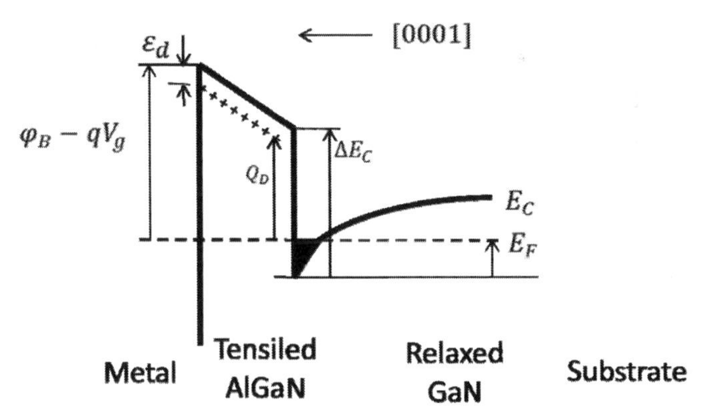

Figure 3.1 Conduction band profile across the AlGaN/GaN heterostructure.

The continuity of the electrical induction $D = \epsilon\epsilon_0 E + P$ across the interface between the $Al_xGa_{1-x}N$ barrier (B) and the GaN channel (C) can be written as

$$\epsilon_0\epsilon_B E_B + P_{sp}^B + P_{pz}^B = \epsilon_0\epsilon_C E_C + P_{sp}^C + P_{pz}^C \tag{3.7}$$

where $\epsilon_0 \approx 8.854 \times 10^{-12}$ Fm^{-1} is the permittivity of a vacuum, $\epsilon_{C,B}$ is the relative permittivity of the channel and the barrier, respectively. The applied voltage V drops in regions B and C as

$$V = E_C W_C + E_B W_B, \tag{3.8}$$

where $W_{B,C}$ is the thickness of barrier and channel, respectively. The GaN layer is relaxed $P^C_{pz} = 0$, so the boundary conditions (3.7) and (3.8) can be solved to find the electric fields:

$$E_B = \frac{\epsilon_C V_g - PW_C/\epsilon_0}{\epsilon_B W_C + \epsilon_C W_B}; \quad E_C = \frac{\epsilon_B V_g - PW_B/\epsilon_0}{\epsilon_B W_C + \epsilon_C W_B}; \tag{3.9}$$

where $P = P^B_{sp} + P^B_{pz} - P^C_{sp}$. If an external bias is absent, $V_g = 0$, the signs of the electric fields depend on the sign of the total polarization difference P which is negative for the structure under consideration at any composition $0 < x \leq 1$. Bearing in mind that the positive [0001] direction points to the left hand side of the picture, one finds that the polarization fields cause the conduction band to decrease in energy with decreasing distance toward the interface from both GaN and AlGaN sides as shown in Fig. 3.1.

Net polarization P creates the positive sheet charge at the AlGaN-barrier side of the interface. The voltage drop across the barrier due to the built-in electric field is

$$U_B = E_B W_B = \frac{-PW_C W_B/\epsilon_0}{\epsilon_B W_C + \epsilon_C W_B} \approx \frac{-PW_B}{\epsilon_0 \epsilon_B} \tag{3.10}$$

If the barrier of area S accumulates a charge Q the capacitance and the sheet charge density are given as

$$C \equiv \frac{Q}{U_B} = \frac{\epsilon_0 \epsilon_B S}{W_B}; \quad \sigma \equiv \frac{CU_B}{S} = -P. \tag{3.11}$$

The charge σ attracts electrons that accumulate in GaN near the interface as illustrated by the shadowed region in Fig. 3.1. These electrons come from the semiconductor GaN layer and compensate the interface charge. In thermodynamic equilibrium the interface electrons create a degenerate electron gas with sheet charge density that can be calculated as the difference between the maximum interface charge σ and the final charge that takes place after equilibrium is reached:

$$qn_S = \sigma - \frac{CU_E}{S}, \tag{3.12}$$

where n_S is the electron sheet density in the channel, U_E is the equilibrium bias across the barrier. The equilibrium bias U_E is less than U_B from Eq. (3.10) due to the partial compensation of the interface charge. It can be found following the energy diagram in Fig. 3.1. Conduction band profile across the AlGaN/GaN heterostructure:

$$qU_E = \varphi_B - (\Delta E_C - E_F) \tag{3.13}$$

The sheet electron density in the channel follows from Eqs. (3.11), (3.12), and (3.13):

$$n_S = \frac{-P}{q} - \frac{\epsilon_0 \epsilon_B}{q^2 W_B}[\varphi_B - qV_g - \Delta E_C + E_F]. \tag{3.14}$$

If the AlGaN barrier is doped, electron transfer occurs from the ionized donors in the barrier to an interface channel. The Fermi energy is to be determined from an additional condition equivalent to the neutrality equation that includes barrier doping. Electron transfer increases the total sheet electron density in the channel

$$N_S = n_S + N_D l, \tag{3.15}$$

where N_D is the volume density of ionized donors and l is the space charge region in the barrier.

It should be noted that in a structure with a thin enough and tensiled AlGaN barrier the space charge region spreads over the whole barrier thickness as illustrated in Fig. 3.1, so l is approximately equal to the barrier thickness. Within simple electrostatic considerations, one may estimate the capacitance and the voltage drop across space charge region as

$$C_D = \frac{\epsilon_0 \epsilon_B S}{l}; \; U_D = \frac{qN_D l^2}{\epsilon_0 \epsilon_B}. \tag{3.16}$$

The energy per donor needed to charge the capacitor of area S is depicted as Q_D in Fig. 3.1:

$$Q_D = \frac{C_D U_D^2}{2} * \frac{1}{lSN_D} = \frac{q^2 N_D l^2}{2\epsilon_0 \epsilon_B}. \tag{3.17}$$

Finally, the energy balance that includes ionized donors can be easily deduced from Fig. 3.1:

$$\Delta E_C = E_F + \varepsilon_D + \frac{q^2 N_D l^2}{2\epsilon_0 \epsilon_B}. \tag{3.18}$$

The total sheet electron density in a two-dimensional channel relates to the Fermi energy as follows:

$$N_S = \frac{mT}{\pi\hbar^2} \text{Log}(1 + \exp[(E_F - \varepsilon_1)/T]), \tag{3.19}$$

where m is the in-plane effective mass of channel electrons, T is the absolute temperature in energy units, ε_1 is the electron ground state energy in the approximately triangular quantum well as illustrated in Fig. 3.1. Equations (3.14), (3.15), (3.18), and (3.19) create a self-consistence set that allows both the polarization and the polarization-induced contribution to the electron density to be calculated.

Polarization doping as wells as built-in electric fields in the barrier and in the channel affect the Rashba spin splitting of channel electrons. The Rashba coefficient and the spin splitting will be discussed below for two important device structures: a modulation-doped heterostructure [8] and a quantum well [9].

3.3 Rashba Interaction in Polarization-Doped Heterostructure

The general expression for the Rashba coefficient (see Chapter 2) being applied to the heterostructure in Fig. 3.1 takes the form

$$\left\langle \frac{\partial \beta}{\partial z} \right\rangle = \Phi^2(0)(\beta_C - \beta_B) + \langle B_B F_B \rangle + \langle B_C F_C \rangle, \tag{3.20}$$

where B and C subscripts stand for the barrier and the channel, respectively, $\langle ... \rangle$ means the average over the ground state wave function in the channel (see Chapter 2).

In the reference frame located at the channel/barrier interface, the electric fields shown in Fig. 3.1 are given as

$$V_C = qF_C z, F_C = \frac{qN_S}{2\epsilon_0\epsilon_C}, z \geq 0;$$

$$V_B = \Delta E_C - qF_B z, F_B = \frac{-P - qN_S}{2\epsilon_0\epsilon_B}, z < 0. \tag{3.21}$$

The potential profile determines the wave function of a confined electron. Ground state wave function can be found by the variational method using the modified Fang–Howard trial function which accounts for the barrier penetration [10]. Electron penetration into the barrier contributes to the spin-orbit coupling parameter (3.20). The trial function has the form

$$\Phi(z) = \begin{cases} \dfrac{Az_0 b^{3/2}}{\sqrt{2}}\exp(\kappa_b z), & z \leq 0, \\ \dfrac{A(z + z_0)b^{3/2}}{\sqrt{2}}\exp(-bz/2), & z > 0, \end{cases}$$

$$\kappa_b = \sqrt{2m_{zB}\Delta E_C / \hbar^2}, \tag{3.22}$$

where b is the variational parameter, z_0 and A follow from normalization and matching conditions for the electron flux at the interface:

$$z_0^{-1} = \frac{\kappa_b m_{zC}}{m_{zB}} + b/2, \quad A^2 = \frac{4z_0^{-2}}{(2z_0^{-1} + b)^2 + b^2 + b^3\kappa_b^{-1}} \tag{3.23}$$

and m_{zB}, m_{zC} are electron masses in the z-direction in the barrier and the channel, respectively.

Total ground state electron energy E_{av} consists of three parts: the kinetic energy, the energy in the channel electric field, and the potential energy induced by other electrons in the channel:

$$E_{av} = \langle T\rangle + \langle V_C\rangle + \frac{1}{2}\langle V_S\rangle, \ \langle T\rangle = \left\langle -\frac{\hbar^2}{2m_{zC}}\frac{\partial^2}{\partial z^2}\right\rangle, \ \langle V_C\rangle = qF_C\langle z\rangle,$$

$$\langle ...\rangle = \int (....)\Phi^2(z)dz, \tag{3.24}$$

The third term in E_{av} is the electron energy in the potential of all other electrons at the interface. The potential can be found as the solution to Poisson's equation

$$V_S = q\varphi,$$

$$\frac{d^2\varphi(z)}{dz^2} = -\frac{qN_S}{\epsilon_0\epsilon_C}\Phi^2(z), \varphi(z=0) = 0, \tag{3.25}$$

where $qN_S\Phi^2(z)$ is the sheet charge density in the channel. The average over the ground state wave function can be expressed as

$$\langle V_S \rangle = \frac{q^2 N_S}{\epsilon_0\epsilon_C}\left\langle z\int_z^\infty \Phi^2(y)dy + \int_0^z y\Phi^2(y)dy \right\rangle. \tag{3.26}$$

The total electron energy in the well E_{av} is subject to minimization with respect to parameter b and $\varepsilon_1 = \min(E_{av})$ is the ground state level in the channel. Parameter b_{min} determines the spatial size of the wave function in the z-direction. Using Eqs. (3.19), (3.21), (3.22), (3.23), and (3.26), one can find the ground state wave function $\Phi(z)$ and energy ε_1, the Fermi energy E_F, and then the total sheet density of electrons in the channel N_S. The Rashba coefficient and the electron spin splitting at the Fermi level can be calculated as $\alpha_R = P_1P_2\left\langle \frac{\partial\beta}{\partial z} \right\rangle$ and $\Delta\varepsilon = 2\alpha_R k$, respectively (see Chapter 2).

Material parameters used in numerical calculations are given in Table 3.1.

To estimate the spin splitting in a barrier with high Al content, we assume that the barrier thickness is equal to 100 Å. The thickness does not exceed the critical value and prevents dislocations from developing in the course of pseudomorphic epitaxial growth up to Al content as high as x = 0.8. This restriction allows using piezoelectric polarization for the pseudomorphic layer, so the partial stress relaxation can be neglected, this otherwise should be taken into account for thicker layers. The barrier doping is chosen as $n = 10^{18}$ cm^{-3}. The ground state wave function and conduction band profile are shown in Fig. 3.2.

The effective thickness of the interface electron spatial distribution is given as $d = \int_0^\infty z\phi^2(z)dz$ and is shown in Fig. 3.3.

Table 3.1 Parameters of the AlGaN material system [8]

Effective mass (m_0)	$0.22 + 0.26x$
Static dielectric constant (ε_0)	$10.4 - 0.3x$
Elastic constants (GPa)	$C_{13} = 103 + 5x$; $C_{33} = 405 - 32x$
Piezoelectric coefficients $\left(\frac{C}{m^2}\right)$	$e_{31} = -0.49 - 0.11x$; $e_{33} = 0.73\,(1 + x)$
Spontaneous polarization $\left(\frac{C}{m^2}\right)$	$P_{\mathrm{sp}} = -0.052x - 0.029$
Piezoelectric polarization $\left(\frac{C}{m^2}\right)$	$P_{\mathrm{PE}} = e_{31}(\varepsilon_{xx} + \varepsilon_{yy}) + e_{33}\varepsilon_{zz}$
Schottky barrier height (eV)	$0.84 + 13x$
Band gap (eV)	$3.4 + 2.7x$
Conduction band offset (eV)	$E_g(x) - E_g(0) - 0.8x$
Lattice constant (Å)	$3.189 - 0.077x$
Donor ionization energy (meV)	38.0
Spin-orbit split energy (meV)	$\Lambda_2 = \Lambda_3 = 6.0$
Crystal-field split energy (meV)	$\Delta_1 = 22.0 - 80.0x$
Interband momentum-matrix elements, $P_{1,2} = \hbar\sqrt{E_{1,2}/2m_0}$	$E_1 = E_2 = 20.0$ eV

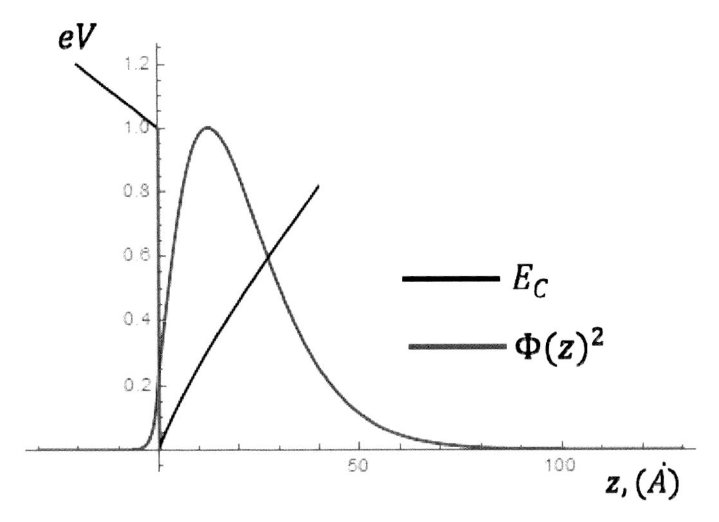

Figure 3.2 Potential profile and ground state wave function in $Al_{0.5}Ga_{0.5}N/GaN$ heterostructure.

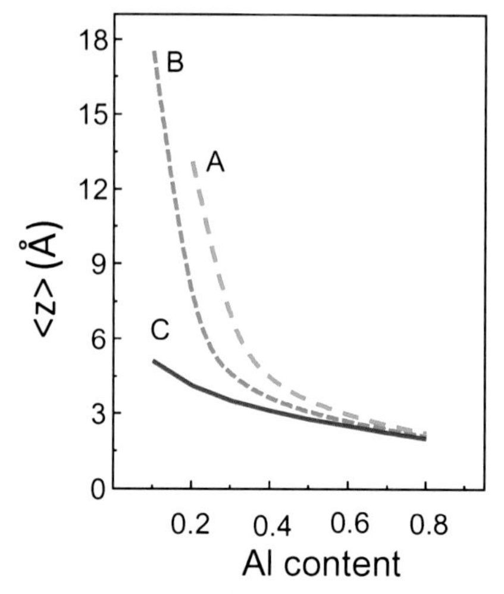

Figure 3.3 Width of the electron channel near GaN/AlGaN interface for various gate voltages. Lines A, B, and C correspond to gate voltages V_g = (−0.8, 0, 0.8) V, respectively.

The positive gate bias decreases the effective channel width, shifting electrons closer to the interface. Quantitative information on the channel width can be useful in managing the magnitude of electron scattering against the interface roughness.

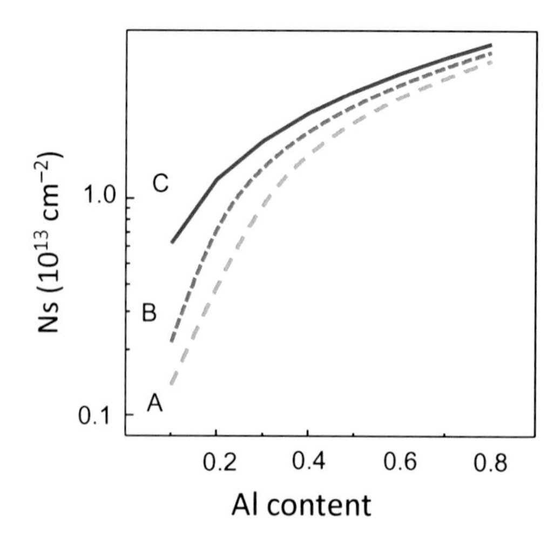

Figure 3.4 Total sheet carrier concentration in a channel with different gate voltages. Lines A, B, and C correspond to gate voltages V_g = (−0.8, 0, 0.8) *V*, respectively.

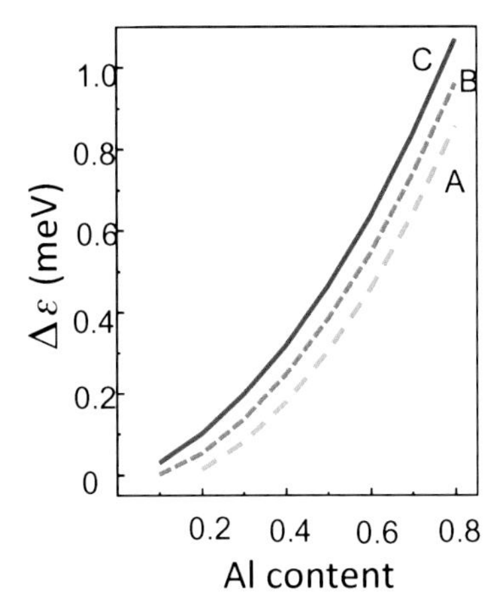

Figure 3.5 Electron spin splitting energy. Lines A, B, and C correspond to gate voltages. V_g (−0.8, 0, 0.8) V, respectively.

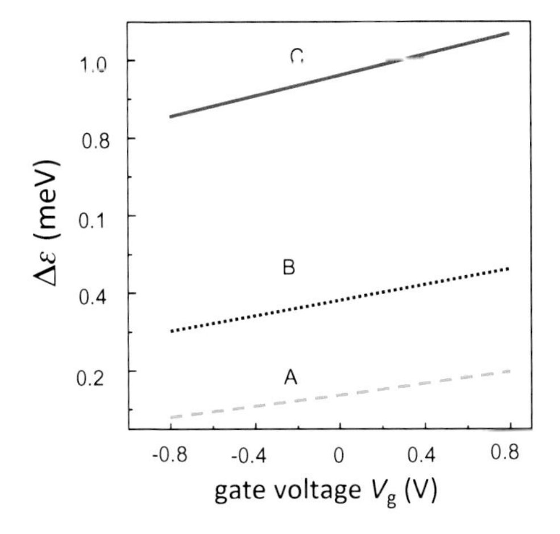

Figure 3.6 Voltage-controlled electron spin splitting. Lines A, B, and C correspond to Al contents x = 0.3, 0.5, 0.8, respectively.

Total sheet electron concentration as a function of Al composition and gate voltage is illustrated in Fig. 3.4.

Figure 3.4 shows how the electron density in the channel can be manipulated by a gate voltage. The built-in electric field in the channel F_C = 1.1 MV/cm at x = 0.2 and F_C = 4.2 MV/cm at x = 0.5, are in a good agreement with those typically observed in GaN/AlGaN QW. Electron spin splitting $|\Delta\varepsilon|$, calculated at the

Fermi level as a functions of Al content and gate bias are shown in Fig. 3.5 and 3.6, respectively.

Spin splitting in narrow-gap III-V materials is in the range of 1–10 meV [1]. The spin splitting, illustrated in Fig. 3.5 and 3.6 is of the same order of magnitude as in narrow-gap III-V materials.

3.4 Structurally Symmetric $In_xGa_{1-x}N$ Quantum Well

The calculations in Section 3.3 were performed for a modulation-doped heterostructure where the near-interface electric field generates Rashba spin splitting whether the polarization field is present or not, so the difference between GaAs and GaN is in numbers only. In a symmetric QW the average electric field and then the Rashba coefficient are zero in [100]-oriented zinc blende structures. However, in a wurtzite QW, spin splitting should exist since polarization fields distort the QW potential profile making the structure electrically asymmetric with finite Rashba spin splitting despite the fact that QW is grown structurally symmetrical. This property of III-nitride structures has been explored in studies of spin injection in AlGaN/GaN heterostructure [11] and topological insulator state in InGaN QW [6]. As the magnitude of the Rashba coefficient relies on polarization fields in III-nitrides, it can be engineered as polarization fields are sensitive to the alloy content and the geometrical parameters of the structure. In this section we outline a simple method that allows us to estimate polarization fields and then calculate the Rashba coefficient in structurally symmetric wurtzite QW [9, 10].

In the reference frame related to a valence band edge of the relaxed InGaN well we can express the strain-induced shift of the valence band following from Eq. (1.34) (Chapter 1) and Eq. (3.5):

$$
\begin{aligned}
V(x) &= (D1 + D3)\,\epsilon_{zz}(x) + 2(D2 + D4)\,\epsilon_{xx}(x) \\
&= 2\left[D2 + D4 - (D1 + D3)\,\frac{C_{13}(x)}{C_{33}(x)}\right]\varepsilon_{xx}(x),
\end{aligned}
$$

$$
\varepsilon_{xx}(x) = \frac{a_{\mathrm{GaN}} - a(x)}{a(x)},\; a(x) = 0.3862x + 3.1986,\ (\text{Å}) \tag{3.27}
$$

Conduction band position is given as

$$C(x) = E_g^0(x) + A_C \varepsilon_{xx}(x), \tag{3.28}$$

where $E_0^g(x) = xE_g^{\mathrm{InN}} + E_g^{\mathrm{GaN}}(1 - x)$ is the bandgap in a relaxed InGaN layer ($\varepsilon_{xx} = 0$), A_C is the conduction band deformation potential. Then the bandgap in a compressed InGaN layer follows from Eqs. (3.27) and (3.28):

$$E_g(x) = C(x) - V(x) \tag{3.29}$$

Once we know the valence band offset $\Delta E_v(x) = 0.26x$ eV [13], the conduction band offset can be estimated as shown in Fig. 3.7:

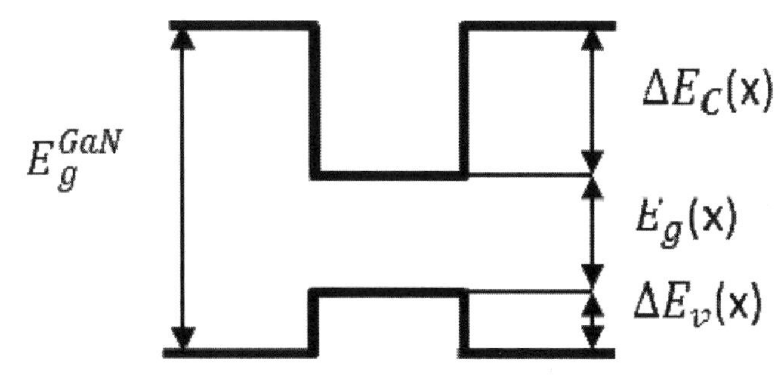

Figure 3.7 Band diagram that relates conduction band offset $\Delta E_C(x)$ to a natural valence band offset $\Delta E_v(x)$.

$$\Delta E_C(x) = E_g^{\mathrm{GaN}} - \Delta E_v(x) - E_g(x) \tag{3.30}$$

To get a band diagram that takes into account built-in electric fields we use In-content dependent total polarization $P = P_{\mathrm{sp}} + P_{\mathrm{pz}}$, where spontaneous P_{sp} and piezoelectric P_{pz} contributions are given as [3]

$$P_{\mathrm{sp}} = 0.042x - 0.034(1 - x) + 0.037x(1 - x),$$

$$P_{\mathrm{pz}} = 0.148x - 0.0424x(1 - x), \left[\frac{C}{m^2}\right]. \tag{3.31}$$

Similar to Eqs. (3.7) and (3.8) we use boundary conditions which are conditions for continuity of electrical induction at two interfaces of three-layer GaN/InGaN/GaN QW along with the distribution of applied voltage across the structure:

$$\begin{cases} \epsilon_L \epsilon_0 \, E_L + P_L = \epsilon_W \epsilon_0 \, E_W + P_W \\ \epsilon_R \epsilon_0 \, E_R + P_R = \epsilon_W \epsilon_0 \, E_W + P_W \\ d_L E_L + d_R E_R + d_W E_W = -V_g, \end{cases} \tag{3.32}$$

where indexes L, R and W stand for left barrier, right barrier, and well, respectively, $P_i(x)$, E_i, and ϵ_i are the total polarization, the electric field, and the dielectric permittivity in layer i, respectively. Piezoelectric polarization is present in the compressed $In_xGa_{1-x}N$ layer and equals zero in the GaN barriers which are assumed to be lattice-matched to a thick GaN buffer layer. Symmetric QW implies barriers of equal thickness, $d_L = d_R = d_B$, and polarizations, $P_R = P_L = P_B$. The solution to Eq. (3.32) is given below:

$$E_L = E_R = E_B = \frac{d_W (P_W - P_B)/\epsilon_0 - \epsilon_W V_g}{2\epsilon_W d_B + \epsilon_B d_W},$$
$$E_W = \frac{2d_B (P_B - P_W)/\epsilon_0 - \epsilon_B V_g}{2\epsilon_W d_B + \epsilon_B d_W}, \tag{3.33}$$

Once we know band offsets $\Delta E_v(x)$, $\Delta E_c(x)$ and internal electric fields $E_W(x)$, $E_B(x)$, the polarization distorted conduction band profile can be determined as shown in Fig. 3.8. The Schrödinger equation solved with this potential profile gives the probability density function also illustrated in Fig. 3.8.

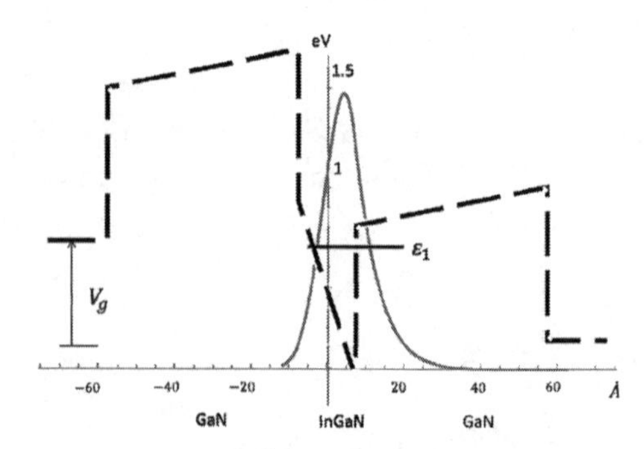

Figure 3.8 Ga-face 50/15/50 Å QW. Potential profile (dashed) and wave function $\Phi^2(z)$ (a.u., solid), $x = 0.5$, $\varepsilon_1 = 0.67$ eV, $V_g = 0.5$ V.

The material parameters used in the numerical calculations are given in Table 3.2 [9].

Table 3.2 Parameters of $In_xGa_{1-x}N$ material system

Effective mass/m_0,	$m_c = 0.2–0.09x$ $m_h = 0.8(1 + x)$
Relative permittivity ϵ	$10.28 + 4.33x$
Band gap in relaxed InGaN (eV)	$E_g(x) = 0.69x + 3.5(1 - x)$
Spin-orbit split energy (meV)	$\Delta_2 = 6.5 - 6.0x$; $\Delta_3 = 13.5 - 9.8x$
Crystal-field split energy (meV)	$\Delta_1 = 21$
Interband momentum-matrix elements	$E_1 = E_2 = 20.0$ eV $P_{1,2} = \hbar\sqrt{E_{1,2}/2m_0}$

The wave function then can be used for the calculation of the effective Rashba coefficient derived in Chapter 2. Below we find the Rashba coefficient and total spin splitting in Ga- and N-face InGaN quantum wells.

3.4.1 Rashba Coefficient in Ga-Face QW

The Rashba coefficient $\alpha_{\text{eff}} = \alpha_{\text{R}} + \alpha_{\text{BIA}}$, $\alpha_{\text{BIA}} = \lambda + \lambda_l \langle k_z^2 \rangle$, where α_{R} is calculated with the expression given in Chapter 2. Numerical calculations have been performed using coefficients values $\lambda = 1.1$ meVÅ, $\lambda_l = 2.4$ eVÅ^3. Figure 3.9 shows α_{R} as a function of the well width.

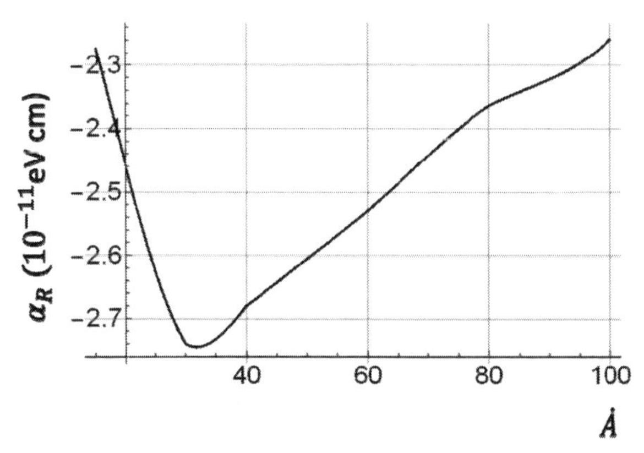

Figure 3.9 Rashba coefficient in Ga-face QW, $x = 0.5$, $d_{\text{L}} = d_{\text{R}} = 75$ Å, $V_g = 0$.

The gate-voltage dependence is shown in Fig. 3.10.

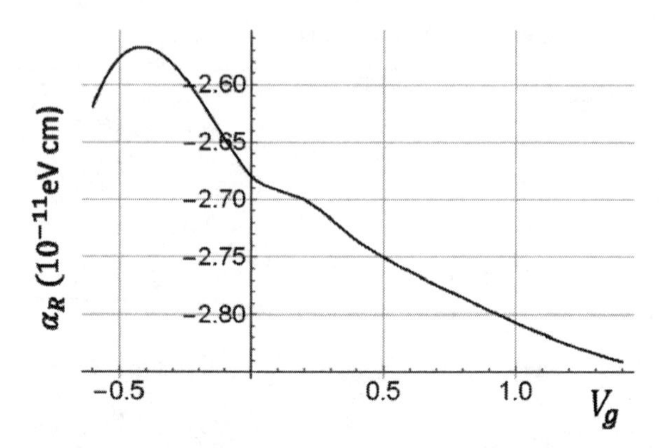

Figure 3.10 Bias dependent Rashba coefficient in Ga-face QW, x = 0.5, d_W = 40 Å, d_L = d_R = 75 Å.

The extent of asymmetry depends on both the well thickness and the applied voltage. Electron confinement in a QW exists in a certain range of gate voltage as the electric field strongly distorts the QW potential profile diminishing the electron confinement. The electric field in the well E_W is shown in Fig. 3.11.

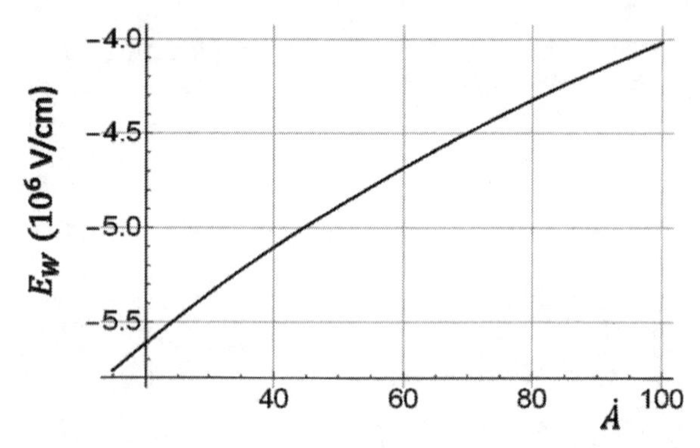

Figure 3.11 Electric field in the well vs. well thickness, x = 0.5, d_L = d_R = 75 Å.

3.4.2 Rashba Coefficient in N-Face QW

Positive E_W is typical for N-face QWs as illustrated in Fig. 3.12. The internal electric fields are opposite to those shown in Fig. 3.8 for Ga-face QWs.

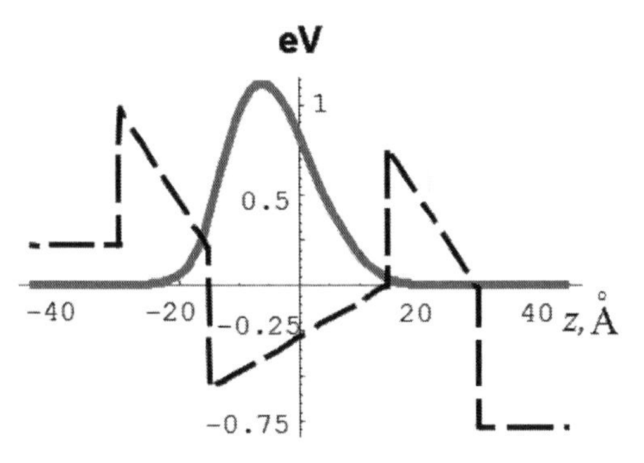

Figure 3.12 Conduction band (eV, dashed) and wavefunction $\Phi^2(z)$ (arbitrary units, solid): *N*-face InGaN/GaN QW; $x = 0.5$, $d_L = d_R =$ 15 Å, $d_W = 30$ Å; $\varepsilon_1 = -0.177$ eV, $V_g = 1$ V.

The Rashba coefficient in an *N*-face QW depends on gate voltage as illustrated in Fig. 3.13.

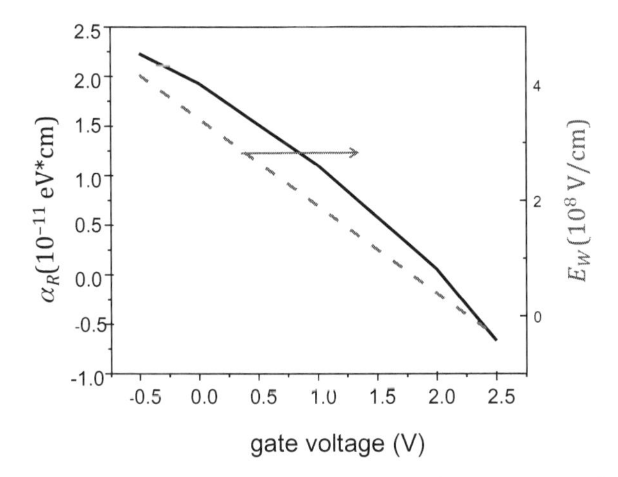

Figure 3.13 Bias dependence of the Rashba coefficient in an *N*-face QW: $x = 0.5$, $d_L = d_R = 15$ Å, $d_W = 30$ Å. Dashed line is the electric field in the well.

As illustrated in Fig. 3.11 and 3.13, the SIA Rashba coefficient is not exactly proportional to the electric field in the well. Instead, a nonlinear relation between them exists.

The overall effective linear spin splitting is determined by an effective Rashba coefficient α_{eff} that also includes BIA contribution

$\alpha_{\rm BIA} = \lambda + \lambda_l \langle k_z^2 \rangle$. The SIA term $\alpha_{\rm R}$ is the most tunable with gate voltage. However, the BIA term constitutes the major contribution to an overall spin splitting: In the quantum wells considered here, the BIA term is 4 to 7 times larger than the Rashba term. Figure 3.14 shows the total spin splitting that includes both SIA and BIA terms in a Ga-face QW.

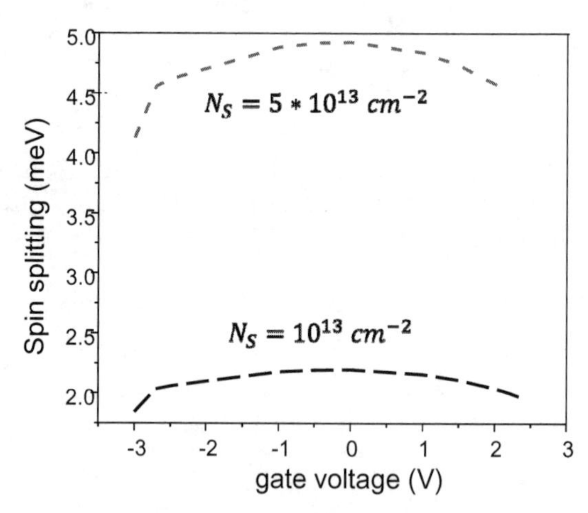

Figure 3.14 Bias dependence of total spin splitting at different doping levels. Ga-face QW: $x = 0.5$, $d_{\rm L} = d_{\rm R} = d_{\rm W} = 50$ Å.

Alloy content, geometry, and gate voltage affect the internal electric field and electron density distribution in the growth direction that has a direct effect on the spin splitting. The SIA spin-orbit coupling coefficients have different signs for Ga-face and N-face III-nitride QWs. The effective linear coupling coefficient is always positive because of the Dresselhaus-type contribution that is dominant in the quantum wells under consideration.

The main point of this chapter is that the SIA Rashba contribution to the overall spin splitting stems from the built-in electric field only as long as one deals with structurally symmetric QWs. However, if structural asymmetry is also present the SIA Rashba term may become larger. The magnitude of the spin splitting estimated in this chapter is comparable with that experimentally observed in III-nitrides and III-V cubic materials.

Electron spin splitting controlled by gate voltage is at heart of the spin transistor proposal [14]. If both the source and the drain are ferromagnetic, the current in the drain depends on the phase difference between spin-up and spin-down electrons and is proportional to $\cos^2[L(k_{F\uparrow} - k_{F\downarrow})]$, L being the channel length. The spin splitting estimated in this chapter can be used for the initial design of the device. Despite the smaller spin-orbit coupling in wide bandgap semiconductors as compared to InGaAs-based III-V materials the overall difference in the Fermi momenta of spin-up and spin-down electrons in III-nitride structures appears to be of the same order of magnitude due to the strong internal polarization fields and polarization doping associated with them.

3.4.3 Inverted Bands in InGaN/GaN Quantum Well

There is a class of materials, topological insulators, which possesses large Rashba coefficient. These materials are mostly Bi_2Te_3-based alloys and their properties will be considered in detail in Chapters 7 and 8. With respect to wide bandgap III-nitrides, it was shown that InN/GaN quantum well also might be in topologically nontrivial state if grown with inverted conduction and valence bands [6]. Band inversion occurs when strong polarization fields push the conduction QW ground energy level lower than the edge of the valence band as illustrated in Fig. 3.15.

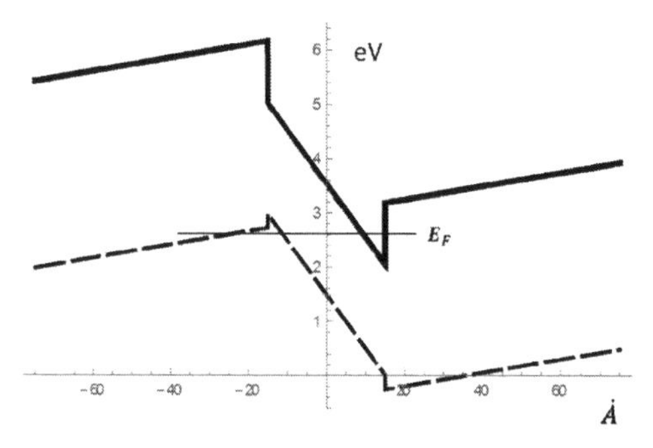

Figure 3.15 Inverted band spectrum in generic GaN/InN/GaN quantum well.

If polarization fields are strong enough and the Fermi energy is located as shown in Fig. 3.15, the half-metallic state exists (both electrons and holes are present at $T = 0$). Conditions for an inverted spectrum depend on an the effective bandgap in the InN well. In the example shown in Fig. 3.15 the bandgap in compressed InN ($\approx$ 2 eV) is larger than that in a relaxed layer $\approx$0.69 eV [5].

3.5 Experimental Rashba Spin Splitting

The magnitudes of Rashba and Dresselhaus coefficients are the key parameters that determine the possibility of using the material system in voltage-controlled spintronic applications. The larger the Rashba spin splitting the higher the spin discrimination that can be achieved; however, the Rashba interaction suppresses the spin lifetime via the Dyakonov–Perel spin relaxation mechanism, so a tradeoff is needed to fit a particular application.

The theoretical calculations of voltage-controlled Rashba splitting in III-nitride modulation-doped structures [8] and quantum wells [9] resulted in a Rashba coefficient of $\approx 2 \times 10^{-13}$ eVm and spin splitting in the 1 to 5 meV range. Experimental confirmation of spin splitting was reported in Ref. [15]. Then the spin splitting and Rashba and Dresselhaus coefficients were measured experimentally in III-nitride heterostructures and quantum wells of various geometry and barrier height. The Dresselhaus coefficient eVm^3 has been reported as 1.6×10^{-31} [16] and 4×10^{-31} [17], and Rashba coefficient (eVm) as 5.5×10^{-13} [16], 6×10^{-13} [18], 2.6×10^{-12} [17, 19], 1.0×10^{-10} [20], 6.8×10^{-11} [21], 7.85×10^{-12} [22], and 4.5×10^{-13} [23]. The Rashba coefficient in bulk wurtzite GaN is less than 4×10^{-15} eVm (see Chapter 2) or at least two orders of magnitude smaller than those observed in quantum wells.

Since electron spin splitting in inversion-asymmetric systems is expected to be exploited in spintronics, the ongoing search for spintronic materials includes new classes of materials such as graphene [24], oxides [25] and BiTe(Se)-based topological insulators (see Chapter 7). Spintronic properties of topological insulators are of special interest as the interplay of Rashba interaction and topologically nontrivial band structure is expected to result in discovery of new spintronic capabilities of materials. In topological insulators Bi_2Se_3 the Rashba coefficient has been found to be $(0.79 - 1.3) \times 10^{-10}$ eVm [26, 27]. Also, a very large Rashba coefficient has been experimentally determined in BiTeJ, $(2 - 4.8) \times 10^{-10}$ eVm [28, 29], BiTeJ [30], and BiTeI/Bi_2Te_3 heterostructures [31].

Problems

3.1 Find coefficients (3.23) from the normalization and matching conditions for the Fang–Howard trial function.

3.2 Solve Poisson's equation (3.25) to obtain the contribution to total electron energy from all other interface electrons, $V_S(z)$.

3.3 Estimate conduction band offset and internal electric fields in 50/50/70 Å Ga-face GaN/$In_{0.3}$ $Ga_{0.7}$ N/GaN QW coherently grown on a thick GaN buffer.

References

1. Awshalom DD, Loss D, Samarth N (2002) *Semiconductor Spintronics and Quantum Computation*, Springer, Berlin, Germany.
2. Loss D, DiVincenzo DP (1998) Quantum computation with quantum dots, *Phys Rev,* **A57**, 120–140.
3. Ambacher O, Majewski J, Miskys C, et al. (2002) Pyroelectric properties of Al(In)GaN/GaN hetero- and quantum well structures, *J. Phys: Condens Mater*, **14**, 3399–3434.
4. Davies RP, Abernathy CR, Pearton SJ (2010) Spintronics in III-nitride based materials, in *Handbook of Spintronic Semiconductors*, Chen WM, Buyanova IA (eds.). pp. 103–121.
5. Walukiewicz W, Li SX, Wu J, Yu KM, Ager III JW, Haller EE, Lu H, Schaff, WJ (2004) Optical properties and electronic structure of InN and In-rich group III-nitride alloys, *J Cryst Growth*, **269**, 119–127.
6. Miao MS, Yan Q, Van de Walle CG, Lou WK, Li LL, Chang K (2012) Polarization-driven topological insulator transition in a GaNInNGaN quantum well, *Phys Rev Lett*, **109**, 186803.
7. Nye JF (1985) *Physical Properties of Crystals: Their Representation by Tensors and Matrices*, Claredon, Oxford.
8. Litvinov VI (2003) Electron spin splitting in polarization-doped group-III nitrides, *Phys Rev,* **B68**, 155314.
9. Litvinov VI (2006) Polarization-induced Rashba spin-orbit coupling in structurally symmetric III-Nitride quantum wells, *Appl Phys Lett,* **89**, 222108.
10. Ando T, Fowler AB, Stern F (1982) Electronic properties of two-dimensional systems, *Rev Mod Phys,* **54**, 437–672; Bastard G (1988) *Wave Mechanics Applied to Semiconductor Heterostructures*, Halsted Press, France.
11. Cu XY, Medvedeva JE, Delley B, Freeman AJ, Stampfl C (2008) Built-in electric field assisted spin injection in Cr and Mn-layer doped AlN/GaN(0001) heterostructures from first principles, *Phys Rev,* **B78**, 245317.

12. Litvinov VI, Manasson, A, Pavlidis D (2004) Short-period intrinsic Stark GaN/AlGaN superlattice as a Bloch oscillator, *Appl Phys Lett*, **85**, 600–602.
13. Wei H, Zunger A (1998) Calculated natural band offsets of all II–VI and III–V semiconductors: Chemical trends and the role of cation *d*-orbitals, *Appl Phys Lett*, **72**, 2011–2013.
14. Datta S, Das B (1990) Electronic analog of the electro-optic modulator, *Appl Phys Lett*, **56**, 665–667.
15. Weber W, Ganichev SD, Danilov SN, et al. (2005) Demonstration of Rashba spin splitting in GaN-based heterostructures, *Appl Phys Lett*, **87**, 262106.
16. Kurdak Ç, Biyikli N, Özgür Ü, Morkoç H, Litvinov VI (2006) Weak antilocalization and zero-field electron spin splitting in $Al_xGa_{1-x}N$/AlN/GaN heterostructures with a polarization-induced two-dimensional electron gas, *Phys Rev*, **B74**, 113308; Cheng H, Biyikli N, Özgür Ü, Kurdak Ç, Morkoç, H, Litvinov VI (2008) Measurement of linear and cubic spin–orbit coupling parameters in AlGaN/AlN/GaN heterostructures with a polarization-induced two-dimensional electron gas, *Physica*, **E40**, 1586–1589.
17. Yin C, Shen B, Zhang Q, Xu F, Tang N, Cen L, Wang X, Chen Y, Yu J (2010) Rashba and Dresselhaus spin-orbit coupling in GaN-based heterostructures probed by the circular photogalvanic effect under uniaxial strain, *Appl Phys Lett*, **97**, 181904.
18. Schmult S, Manfra MJ, Punnoose A, Sergent AM, Baldwin KW, Molnar R (2006) Large Bychkov-Rashba spin-orbit coupling in high-mobility GaN/Al_xGa_{1-x} N heterostructures, *Phys Rev*, **B74**, 033302.
19. Cho KS, Liang CT, Chen YF, Fan JC (2007) Demonstration of Rashba spin splitting in an $Al_{0.25}Ga_{0.75}$ N/GaN heterostructure by microwave-modulated Shubnikov–de Haas oscillations, *Semicond Sci Technol*, **22**, 870–874; Spirito D. Frucci G. Di Gaspare A. Di Gaspare L. Giovine E. Notargiacomo A. Roddaro S. Beltram F. Evangelisti F. (2011) Quantum transport in low-dimensional AlGaN/GaN system, *J Nanopart Res*, **13**, 5699–5704.
20. Belyaev AF, Raicheva VG, Kurakin AM, Klein N, Vitusevich SA (2008) Investigation of spin-orbit interaction in AlGaN/GaN heterostructures with large electron density, *Phys Rev*, **B77**, 035311.
21. Zhou WZ, Lin T, Shang LY, Sun L, Gao KH, Zhou YM, Yu G, Tang N, Han K, Shen B, Guo SL, Gui YS, Chu JH (2008) Weak antilocalization and beating pattern in high electron mobility Al_xGa_{1-x} N/GaN two-dimensional electron gas with strong Rashba spin-orbit coupling, *J Appl Phys*, **104**(5), 053703.

22. Lisesivdin SB, Balkan N, Makarovsky O, et al. (2009) Large zero-field spin splitting in AlGaN/AlN/GaN/AlN heterostructures, *J Appl Phys*, **105**, 093701.

23. Stefanowicz W, Adhikari R, Andrearczyk T, Faina B, Sawicki M, Majewski JA, Dietl T, Bonanni A (2014), Experimental determination of Rashba spin-orbit coupling in wurtzite *n*-GaN:Si, *Phys Rev*, **B89**, 205201.

24. Varykhalov A, Sanchez-Barriga J, Shikin AM, Biswas C, Vescovo E, Rybkin A, Marchenko D, Rader O (2008) Electronic and magnetic properties of quasifree standing graphene on Ni, *Phys Rev Lett*, **101**, 157601.

25. Kozuka Y, Teraoka S, Falson J, Oiwa A, Tsukazaki A, Tarucha S, Kawasaki M (2013) Rashba spin-orbit interaction in a $Mg_xZn_{1-x}O$/ZnO two-dimensional electron gas studied by electrically detected electron spin resonance, *Phys Rev*, **B87**, 205411.

26. King PDC, Hatch RC, Bianchi M, Ovsyannikov R, Lupulescu C, Landolt G, Slomski B, Dil JH, Guan D, Mi JL, Rienks EDL, Fink J, Lindblad A, Svensson S, Bao S, Balakrishnan G, Iversen BB, Osterwalder J, Eberhardt W, Baumberger F, Hofmann Ph (2011) Large tunable Rashba spin splitting of a two-dimensional electron gas in Bi_2Se_3, *Phys Rev Lett*, **107**, 096802.

27. Zhu ZH, Levy G, Ludbrook B, Veenstra CN, Rosen JA, Comin R, Wong D, Dosanjh P, Ubaldini A, Syers P, Butch NP, Paglione J, Elfimov IS, Damascelli A (2011) Rashba spin-splitting control at the surface of the topological insulator Bi_2Se_3, *Phys Rev Lett*, **107**, 186405.

28. Eremeev SV, Nechaev IA, Koroteev Yu M, Echenique PM, Chulkov EV (2012) Ideal two-dimensional electron systems with a giant Rashba-type spin splitting in real materials: Surfaces of Bismuth Tellurohalides, *Phys Rev Lett*, **108**, 246802.

29. Rusinov P, Nechaev IA, Eremeev SV, Friedrich C, Blügel S, Chulkov EV (2013) Many-body effects on the Rashba-type spin splitting in bulk bismuth tellurohalides, *Phys Rev*, **B87**, 205103.

30. Ishizaka K, Bahramy MS, Murakawa H, Sakano M, Shimojima T, Sonobe T, Koizumi K, Shin S, Miyahara H, Kimura A, Miyamoto K, Okuda T, Namatame H, Taniguchi M, Arita R, Nagaosa N, Kobayashi K, Murakami Y, Kumai R, Kaneko Y, Onose Y, Tokura Y (2011) Giant Rashba-type spin splitting in bulk BiTeI, *Nat Mater*, **10**, 521–526.

31. Zhou JJ, Feng W, Zhang Y, Yang SA, Yao Y (2014) Engineering topological surface states and giant Rashba spin splitting in BiTeI/Bi_2Te_3 heterostructures, *Sci Rep*, **4**, 3841.

Chapter 4

Tunnel Spin Filter in Rashba Quantum Structure

Spin-dependent electron tunneling in III-V semiconductors has become the subject of intensive study in the search for an effective non-magnetic spin injector. Non-magnetic spin filters are appealing since they do not require a magnetic field and they allow spin manipulation to be achieved with applied voltage. It is known that the direct contact between a ferromagnetic metal and a semiconductor has low spin injection efficiency due to the conductivity mismatch [1]. To improve the efficiency of spin transfer between a metal and a semiconductor, electrical contacts of increased resistance are proposed in Ref. [2]. The possible implementation of this approach is tunnel contact between nonmagnetic semiconductors with intrinsic spin splitting. Spin selection in tunnel contacts occurs in the course of electron transmission through a single- or double-barrier-quantum well (QW) if the spin-orbit interaction is essential. The study of tunnel spin discrimination started with cubic [100]-oriented InGaAs-based heterostructures. The case of wurtzite AlGaN quantum wells is more complex as symmetry allows intrinsic linear k-dependent spin splitting to exist. The polarization fields distort the band structure making the extrinsic Rashba term nonzero as the structure becomes electrically asymmetric. One more difference between cubic and wurtzite structures is that the Dresselhaus and Rashba terms in

Wide Bandgap Semiconductor Spintronics
Vladimir Litvinov

ISBN 978-981-4669-70-2 (Hardcover), 978-981-4669-71-9 (eBook)
www.panstanford.com

wurtzite structures have the same momentum-dependent spin symmetry (see Chapter 2).

In wide bandgap III-nitrides, the spin injection efficiency is affected by polarization fields. In Chapter 3 we discussed spin splitting and concluded that it has a magnitude comparable to that of the GaAs counterparts. In this chapter we deal with voltage-controlled tunneling spin injection in Al(In)GaN/GaN QW. We study spin tunneling and the role strain plays in spin injection. With respect to tunneling, it is useful to note the essential difference between the two types of QWs: InGaAs-based [001]-oriented ZB and GaN-based [0001]-wurtzite (W). The voltage applied across the ZB structure induces electric fields of the same sign in each layer and the flat-band approximation well describes the electron band edges if the external bias becomes zero. In a W-QW the built-in electric fields, caused by spontaneous and lattice-mismatch piezoelectric polarizations distort conduction and valence band profiles, causing Zener tunneling processes to occur, even if an external bias is not applied.

In Section 4.1, the tunnel spin polarization in a piezoelectric AlGaN/GaN double barrier structure is calculated. The built-in fields make a difference in the voltage-controlled spin tunneling, namely, the spontaneous and piezoelectrical polarization fields increase spin polarization efficiency that otherwise would be low if no electrical polarization were taken into account. The relation between the polarization and the spin orientation allows engineering the spin injection by varying the Al-content in barriers and so manipulating the lattice-mismatch strain.

Section 4.2 deals with spin-polarized tunnel transmission through a single barrier. The transmission and its bias dependence are calculated for electrons that experience interface elastic and spin-orbit impurity scattering. The spin-orbit part of impurity scattering enhances the spin polarization of transmitted electrons and over-barrier reflection results in a resonant spin filtering.

4.1 Double-Barrier Resonant Tunneling Diode

4.1.1 Current–Voltage Characteristics

The total current which flows in the z-direction through a generic tunneling structure can be derived by calculating the difference between emitter-collector flows in opposite directions: $J = J_{e\to c} - J_{c\to e}$. The emitter-collector current density can be written as

$$J_{e\to c} = q\hbar^3 \int v_z(\mathbf{k})T(\mathbf{k})[1 - R(\mathbf{k})]\, n_e(\mathbf{k})d\mathbf{k} = \frac{q}{(2\pi)^3}\int d\mathbf{k}_{\|} \int dE_z T(E)[1 - R(E)] f_e$$

$$v_z = \frac{1}{\hbar}\frac{\partial E_z}{\partial k_z};\ n_e(\mathbf{k}) = \frac{1}{h^3} f_e;\ f_e = \frac{1}{1 + \exp[(E - E_F)/k_B T])}, \tag{4.1}$$

where $\mathbf{k}$ is the three-dimensional wave vector, $\mathbf{k}_{||}$ is the in-plane wave vector, E is the energy of an incoming electron, $E = \hbar^2 k^2/2m = E_x + E_y + E_z$, T is the emitter-collector transmission probability, R is the reflection probability in the collector, E_F is the Fermi energy, $n_e(\mathbf{k})$ is the number of tunneling electrons of a particular spin orientation per unit volume in k-space. Reflection probability from the collector is the electron fill-factor

$$R(E) = f_c = \frac{1}{1 + \exp[(E + qV - E_F)/k_B T]}, \tag{4.2}$$

where V is the external bias applied to the collector region. The resulting current is given as

$$J_{e\to c} = \frac{q}{(2\pi)^3 \hbar}\int d\mathbf{k}_{\|} dE_z T(E) f_c f_e \exp[(E + qV - E_F)/k_B T]. \tag{4.3}$$

The current in the opposite direction is obtained by switching the expressions for f_c and f_e. Then the total current reads $J = J_{e\to c} - J_{c\to e}$ and, after integration over the in-plane momentum direction, takes the form

$$J = \frac{mqk_B T}{4\pi^2\hbar^3}\int_0^\infty dE \int_0^{y(E)} dy\, T(E, y)\Phi(E, y),$$

$$\Phi = \frac{\exp(v) - 1}{1 + \exp(-x - y) + \exp(x + y + v) + \exp(v)},\ x = \frac{E - E_F}{k_B T},\ y = \frac{\hbar^2 k_{\|}^2}{2mk_B T},\ v = \frac{qV}{k_B T}, \tag{4.4}$$

The momentum integration limit $y(E)$ is determined by the condition for both incident and outgoing electron momenta to be real. The transmission coefficient is expressible as

$$T = Tr(tt^{+}) = (|t_{+}|^{2} + |t_{-}|^{2}),\ t = \begin{pmatrix} t_{+} & 0 \\ 0 & t_{-} \end{pmatrix}, \tag{4.5}$$

where $t_{\pm}$ are the complex transmission amplitudes for spin-up and spin-down electrons, respectively. Total electron current Eq. (4.4) can be represented as a sum of electron fluxes in $\pm$ spin states and presents the current voltage characteristic of the tunnel junction.

4.1.2 Spin Current

The electric current quantitatively describes the charge transfer; it obeys the particle conservation law and becomes zero in thermodynamic equilibrium when applied voltage is absent. The concept of spin current is more complicated as the spin current is not conserved in systems with spin-orbit interaction. This non-conservation results in a permanent spin flow in equilibrium. Permanent spin current, however, does not produce any observable effect such as spin accumulation or transfer of a finite spin density from one part of the structure to another.

Below the spin current is treated as a pseudotensor, as defined in Ref. [3]. The spin flux in the z-direction is the axial vector that determines the direction and magnitude of spin polarization:

$$\mathbf{J}_{s} = \frac{1}{(2\pi)^{3}\hbar}\int (f_{e} - f_{c})\mathbf{T}(E, \mathbf{k}_{\parallel})dE_{z}d\mathbf{k}_{\parallel},$$

$$\mathbf{T} = Tr(t\mathbf{s}t^{+}),\ \mathbf{s} = \begin{pmatrix} \mathbf{S}_{+} & 0 \\ 0 & \mathbf{S}_{-} \end{pmatrix},\ \mathbf{S}_{\pm} = -\mathbf{S}_{\mp} = \frac{1}{2} < u_{\pm}|\sigma|u_{\pm} >. \tag{4.6}$$

where $\alpha_{x,y,z}$ are the Pauli matrices, $u_{\pm}$ are the spinors for up- and down-spin states, respectively.

The tunneling amplitude through the Rashba electron gas preserves the symmetry $t_{\uparrow}(\mathbf{k}_{||}) = t_{\downarrow}(-\mathbf{k}_{||})$ that originates from the Kramers degeneracy of electron spectrum in the quantum well. As long as the distribution functions are the equilibrium in-plane Fermi functions, $f^{0}_{e,c}(\mathbf{k}_{||}) = f^{0}_{e,c}(-\mathbf{k}_{||})$, the spin current (4.6) equals zero due to the symmetry of both the in-plane electron spectrum and transmission amplitudes. So, the vertical bias that forces the electric current to flow across the structure does not cause the spin polarization in neither emitter nor collector regions.

The lateral electric field **F**, applied to the collector region, breaks the symmetry by introducing non-equilibrium correction to the distribution function $f_c = f_c^0 + f', f' = \frac{\hbar q \tau}{m_{||}} \frac{\partial f_c^0}{\partial E} \mathbf{k}_{||} \mathbf{F}$, τ is the momentum relaxation time. Then, Eq. (4.6) can be rewritten in the form

$$\mathbf{J}_s = -\frac{q\tau(2m_{||}k_BT)^{\frac{3}{2}}}{32\pi^2\hbar^3 m_{||}} \int_0^{\infty} dE_z \int_0^{y_{max}} dy\sqrt{y}\,\frac{\partial f_c^0}{\partial E}[|t_\uparrow|^2 - |t_\downarrow|^2][\mathbf{n}\times\mathbf{F}], \quad (4.7)$$

where **n** is the unit vector perpendicular to the QW interface as illustrated in Fig. 4.1.

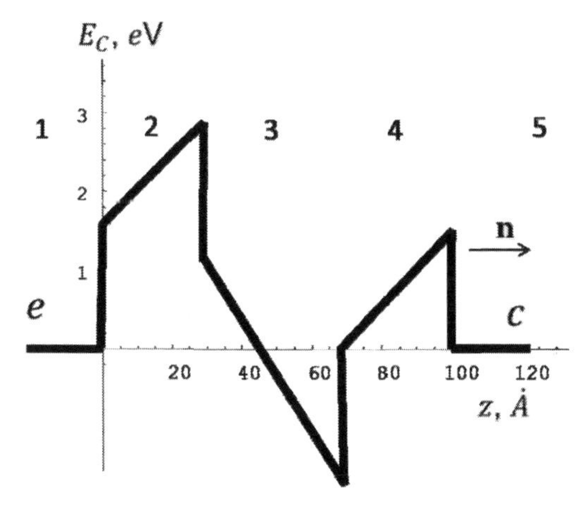

Figure 4.1 Conduction band profile of an unbiased AlGaN/GaN/AlGaN double barrier structure; regions 1 and 5 are the emitter and collector, respectively.

Single-mode tunneling at a particular in-plane momentum can be characterized with the spin-selective transparency $T_\pm = |t_\pm|^2$ and the tunnel spin polarization $p = \frac{T_+ - T_-}{T_+ + T_-}$. The single-mode tunnel spin polarization can be as high as 100% at a resonance; however, this does not mean that the effective spin injection occurs when the tunnel current flows across the structure. The total spin flux contains all possible in-plane modes, weighted with the equilibrium distribution functions, and the most important modes are those close in energy to the Fermi level in the collector region to which the lateral electric field is applied. So, overall spin injection cannot be adequately represented by the single-mode spin polarization P, and the magnitude of spin current J_S should be used instead as it contains integration over all in-plane modes.

Equations (4.4) and (4.7) present the vertical I–V characteristics and the spin polarization current, respectively. It should be noted that the lateral electric field-induced non-equilibrium correction to the distribution function does not contribute to the current–voltage characteristics Eq. (4.4) as the non-equilibrium term becomes zero after integration over the in-plane momentum directions. So, the spin current is proportional to the lateral electric field and it cannot be expressed merely as the difference of the vertical charge currents J calculated separately for $\pm$ electron states.

The spin polarization resulting from tunneling has the in-plane direction perpendicular to the lateral electric field $\mathbf{J_S} \sim [\mathbf{n} \times \mathbf{F_L}]$. From the phenomenological standpoint, the effect is similar to the in-plane-current induced spin orientation in an electron gas with the k-linear spin splitting [4] or in Rashba two-dimensional electron gas [5, 6]. What discerns the spin injection considered here, from systems described in Refs. [4–6], is that the spin polarization appears in the bulk collector region where the in-plane Rashba spin splitting is absent; the spin imbalance comes from the vertical tunneling that provides the spin injection into an active region of the spin field-effect transistor. Device implementation of the vertical tunnel spin injector is proposed in Ref. [7].

4.1.3 Tunnel Transparency

The electron Hamiltonian in each [0001]-oriented layer includes kinetic energy and spin-orbit interaction (see Chapter 2):

$$H = \frac{\hbar^2 k^2}{2m} + k_z\left[\frac{\hbar^2}{2m_z} + \lambda_l(\sigma_x k_y - \sigma_y k_x)\right]k_z + \lambda_t(\sigma_x k_y - \sigma_y k_x)k^2$$
$$+\left(\lambda + \frac{1}{2}\frac{\partial\beta}{\partial z} + \frac{\beta}{2}\frac{\partial}{\partial z}\right)(\sigma_x k_y - \sigma_y k_x) + V(z), \tag{4.8}$$

where β includes matrix elements $P_{1,2}$ and relates to the Rashba coupling parameter, λ is the bulk linear spin-orbit coupling constant, $\lambda_{l,t}$ are the Dresselhaus spin-orbit interaction constants, $m_{||}$, m_z are the effective masses, $k_z = -i\frac{\partial}{\partial z}$, $V(z)$ is the heterostructure potential energy that includes an external bias V_{ext}. The symmetric Rashba term in (4.8) should be used in order to derive the correct boundary conditions at each interface as shown below. The coupling constant λ_t in Eq. (4.8) renormalizes the in-plane effective mass and will be neglected.

After the unitary transformation from the initial z-oriented spinors to a new basis $u_{\pm} = U_W|\uparrow\downarrow> = \frac{1}{\sqrt{2}}\begin{pmatrix} \mp ie^{-i\varphi} \\ 1 \end{pmatrix}$, the Hamiltonian becomes diagonal as expressed below:

$$H = \frac{\hbar^2 k^2}{2m} + k_z \frac{\hbar^2}{2m_z^{\pm}} k_z \mp k\left(\lambda + \frac{1}{2}\frac{\partial \beta}{\partial z} + \frac{\beta}{2}\frac{\partial}{\partial z}\right) + V(z),$$

$$m_z^{\pm} = m_z\left(1 \mp \frac{2m_z k}{\hbar^2}\right)^{-1}. \tag{4.9}$$

Tunneling structure shown in Fig. 4.1 comprises five regions: Thick emitter and collector GaN leads (regions 1 and 5) and GaN QW (region 3) are placed between two GaN barriers (regions 2 and 4). Since the whole structure is lattice-matched to GaN, we account for the tensile strain in the barriers. The corresponding conduction band offsets, polarization fields (spontaneous and piezoelectric), and total internal electric fields F_j in each layer of thickness GaN can be calculated as described in Chapter 3 (see details in Ref. [8]). The resulting conduction band profile is shown in Fig. 4.1.

Built-in electric fields are specific to GaN-based structures and when constructing the wave functions in the well and the barriers where electric fields are strong, instead of plane waves, one has to choose a basic set of Airy functions $Ai(z)$, $Bi(z)$, that is electron eigenfunctions in the presence of an electric field. Still, plane waves may serve as a basis in thick emitter (1) and collector (5) regions shown in Fig. 4.1. Electron wave functions in layers, numbered in Fig. 4.1, can be represented as follows:

$$\Phi_{j\pm} = u_{\pm}\Psi_{j\pm}(z)\exp(i\mathbf{k}_{||}\mathbf{r}_{||}), \Psi_{1\pm}(z) = A_{1\pm}\exp(ik_1 z) + B_{1\pm}\exp(-ik_1 z),$$
$$\Psi_{2\pm}(z) = A_{2\pm}Ai(\rho_{2\pm}) + B_{2\pm}Bi(\rho_{2\pm}), \Psi_{3\pm}(z) = A_{3\pm}Ai(\rho_{3\pm}) + B_{3\pm}Bi(\rho_{3\pm}),$$
$$\Psi_{4\pm}(z) = A_{4\pm}Ai(\rho_{4\pm}) + B_{4\pm}Bi(\rho_{4\pm}), \Psi_{5\pm}(z) = A_{5\pm}\exp(ik_5(z - d_2 - d_3 - d_4), \tag{4.10}$$

where $r_\pm$, $t_\pm$ are the reflection and transmission amplitudes, respectively, and

$$\frac{\hbar k_1}{\sqrt{2m_z}} = \sqrt{E - \frac{\hbar^2 k^2}{2m_{||}}}, \qquad \frac{\hbar k_5}{2m_z} = \sqrt{E + qV_{\text{ext}} - \frac{\hbar^2 k^2}{2m_{||}}},$$
$$\rho_{2\pm}(z) = C_{2\pm}(qF_2 z + \Delta E_{c1} - E), \ \rho_{3\pm}(z) = C_{3\pm}[qF_3(z - d_2) + qF_2 d_2 - E],$$
$$\rho_{4\pm}(z) = C_{4\pm}[qF_4(z - d_2 - d_3) + qF_3 d_3 + qF_2 d_2 + \Delta E_{c2} - E],$$
$$C_{j\pm} = \left(\frac{2m_{jz}^{\pm}}{\hbar^2 q^2 F_j^2}\right)^{1/3}, \tag{4.11}$$

where $\Delta E_{c1,2}$ are conduction band offsets on interfaces 2/3 and 3/4, respectively. The electric fields in Eq. (4.11) carry appropriate signs as illustrated in Fig. 4.1: $F_{2,4} > 0$, $F_3 < 0$.

The tunnel transparency of the structure can be found using the transfer matrix method. First, it is necessary to derive the boundary conditions corresponding to the Hamiltonian (4.9). At the interface $z = 0$ the first boundary condition follows from matching wave functions at the interface between the left (L) and right (R) regions. Another condition we obtain by integrating the Schrödinger equation $H_\pm \Psi_\pm = E_\pm \Psi_\pm$ over the infinitesimal region across the interface:

$$\Psi_{L\pm}(0) = \Psi_{R\pm}(0),$$
$$\frac{\hbar^2}{2m_z^{\pm}(L)} \frac{\partial \Psi_{L\pm}}{\partial z}\bigg|_{z=0} - \frac{\hbar^2}{2m_z^{\pm}(R)} \frac{\partial \Psi_{R\pm}}{\partial z}\bigg|_{z=0} \mp \frac{1}{2} k \Psi_{\pm}(0)(\beta_R - \beta_L) = 0. \tag{4.12}$$

Writing down boundary conditions (4.12) for all four interfaces one obtains a set of equations for coefficients $A_{j\pm}$ and $B_{j\pm}$. At each interface these coefficients relate through the transfer matrix. Generic boundary conditions (4.12) applied to the 1/2 interface take the form

$$A_{1\pm} + B_{1\pm} = A_{2\pm} Ai(\rho_{2\pm}(0)) + B_{2\pm} Bi(\rho_{2\pm}(0)),$$
$$\frac{\hbar^2}{2m_1} ik_1(A_{1\pm} - B_{1\pm}) - \frac{\hbar^2}{2m_2} \rho'(0)[A_{2\pm} Ai'(\rho_{2\pm}(0)) + B_{2\pm} Bi'(\rho_{2\pm}(0))]$$
$$\mp \frac{\beta_2 k}{2} [A_{2\pm} Ai(\rho_{2\pm}(0)) + B_{2\pm} Bi_{2\pm}(\rho_{2\pm}(0))] \pm \frac{\beta_1 k}{2} [A_{1\pm} + B_{1\pm}] = 0, \tag{4.13}$$

where $\rho' = \partial \varrho/\partial z|_{z=0}$; $\partial Ai(u)/\partial u|_{u=a}$.

In matrix form Eq. (4.13) reads

$$M_1\begin{pmatrix}A_{1\pm}\\B_{1\pm}\end{pmatrix} = M_2(0)\begin{pmatrix}A_{2\pm}\\B_{2\pm}\end{pmatrix},$$

$$M_1 = \begin{pmatrix}1 & 1\\ \frac{i\hbar^2 k}{2m_1} \pm \frac{\beta_1 k}{2} & -\frac{i\hbar^2 k}{2m_1} \pm \frac{\beta_1 k}{2}\end{pmatrix},$$

$$M_2(0) = \begin{pmatrix}Ai(\rho_{2\pm}(0)) & Bi(\rho_{2\pm}(0))\\ \frac{\hbar^2}{2m_2}\rho'(0)Ai'(\rho_{2\pm}(0)) \pm \frac{\beta_2 k}{2}Ai(\rho_{2\pm}(0)) & \frac{\hbar^2}{2m_2}\rho'(0)Bi'(\rho_{2\pm}(0)) \pm \frac{\beta_2 k}{2}Bi(\rho_{2\pm}(0))\end{pmatrix}. \quad (4.14)$$

Equation (4.14) gives the transfer matrix $R_{12}(0)$ that relates the coefficients at interface $1/2(z = 0)$:

$$\begin{pmatrix}A_{1\pm}\\B_{1\pm}\end{pmatrix} = R_{12}^{\pm}(0)\begin{pmatrix}A_{2\pm}\\B_{2\pm}\end{pmatrix}, \quad R_{12}^{\pm}(0) = M_1^{-1}M_2(0). \quad (4.15)$$

Similar relations at interfaces 2/3, 3/4, and 4/5 which are obtained the same way result in relations at interfaces located at $z = d_2$, $d_2 + d_3$, and $d_2 + d_3 + d_4$

$$R_{23}^{\pm}(d_2) = [M_2(d_2)]^{-1}M_3(d_2),$$
$$R_{34}^{\pm}(d_2 + d_3) = [M_3(d_2 + d_3)]^{-1}\, M_4(d_2 + d_3),$$
$$R_{45}^{\pm}(d_2 + d_3 + d_4) = [M_4(d_2 + d_3 + d_4)]^{-1}\, M_5, \quad (4.16)$$

Then the relation between incoming and outgoing amplitudes is expressed as

$$\begin{pmatrix}A_{1\pm}\\B_{1\pm}\end{pmatrix} = \mathrm{TM}\begin{pmatrix}A_{5\pm}\\B_{5\pm}\end{pmatrix}, \quad (4.17)$$

where the total transfer matrix can be written as

$$\mathrm{TM} = R_{12}^{\pm}(0)R_{23}^{\pm}(d_2)R_{34}^{\pm}(d_2 + d_3)R_{45}^{\pm}(d_2 + d_3 + d_4). \quad (4.18)$$

Let us consider tunneling from region 1 to region 5 under the assumption of a unit amplitude incoming wave, $A_1 = 1$, and zero reflected amplitude in region 5, $B_5 = 0$. The amplitude of the transmitted wave follows from Eq. (4.17):

$$A_5 = \frac{1}{[\mathrm{TM}]_{11}}. \tag{4.19}$$

Electron flux

$$j = \frac{i\hbar}{m}[\Psi(z)\nabla\Psi^*(z) - \Psi^*(z)\nabla\Psi(z)], \tag{4.20}$$

calculated with $\Psi_1(z)$ and $\Psi_5(z)$ gives us the transmission coefficient defined as the ratio of the electron flux in the outgoing wave to that in the incoming wave:

$$T = \left|\frac{1}{[\mathrm{TM}]_{11}}\right|^2 \frac{k_5}{k_1} \tag{4.21}$$

Transmission exists in the region of in-plane wave vectors where both k_1 and k_5 are real numbers.

4.1.4 Polarization Fields

Tunnel transparency is sensitive to the spatial profile of the potential barrier. For the double barrier structure shown in Fig. 4.1, the polarization fields in layers 2, 3, and 4 shape the potential profile and along with the conduction band offset determine the efficiency of spin filtering.

In order to estimate the conduction band offset and polarization fields we use the method applied to InGaN QW in Chapter 3. In an $Al_xGa_{1-x}N$ structure the natural valence band offset of 0.8 eV between GaN and AIN [9] can be linearly interpolated for any alloy composition as $\Delta E_v(x) = 0.8x$. The valence band edge in GaN is chosen as the reference energy. Also we assume linear composition dependence (Vegard's law) of the inplane lattice constant, the effective mass in the z-direction, and the static dielectric constant:

$$a(x) = xa_{\mathrm{AlN}} + (1-x)a_{\mathrm{GaN}};\ m_z(x) = xm_{\mathrm{AlN}} + (1-x)m_{\mathrm{GaN}},$$
$$\epsilon_p(x) = 10.28 + 0.03x. \tag{4.22}$$

The bandgap in a relaxed bulk crystal has nonlinear alloy dependence with the bowing parameter of 1 eV:

$$E_g(x) = xE_{g\mathrm{AlN}} + (1-x)\,E_{g\mathrm{GaN}} + x(1-x). \tag{4.23}$$

If the structure is coherently grown on a GaN buffer then the AlGaN barrier layers are under tensile strain $\varepsilon(x) = [a(0) - a(x)]/(a(0))$. Strain shifts the conduction (E_C) and valence band (E_V) edges, so in strained layers $E_g(x)$ is no longer a real bandgap. The band edges shift as follows

$$\begin{aligned} E_c(x) &= E_g(x) + 3A_c\varepsilon(x), \\ E_v(x) &= 2\varepsilon(x)\left[D_2 + D_4 - (D_1 + D_3)\frac{C_{11}(x)}{C_{33}(x)}\right], \end{aligned} \tag{4.24}$$

where C_{ij} and A_c, D_i are the elastic constants and deformation potentials, respectively. Deformation potentials were introduced in Chapters 1 and 3. Their numerical values and composition dependence can be found in Refs. [10, 11].

The observable bandgap and thus the conduction band offset become

$$\begin{aligned} E_G(x) &= E_c(x) - E_v(x). \\ \Delta E_c(x) &= E_G(x) - E_G(0) - \Delta E_v(x) \end{aligned} \tag{4.25}$$

To render the correct band diagram in Fig. 4.1, we have to account for all sources of the electric field: spontaneous and piezoelectric polarizations, $P(x) = P_{\mathrm{sp}}(x) + P_{\mathrm{pz}}(x)$ as well as an external field in a biased structure. In what follows we assume that barriers in Fig. 4.1 have different alloy compositions, x in barrier 2 and y in barrier 4. For a multilayer structure biased with external voltage V we deal with the continuity of the electrical induction $D = \epsilon_{\mathrm{p}}\epsilon_0 F + P$ across the interfaces, where F is an electric field to be determined. This gives us the set of equations

$$\begin{cases} F_2 d_1 + F_3 d_2 + F_4 d_3 = -V, \\ \epsilon_0\ \epsilon_p(x) F_2 + P_2(x) = \epsilon_0\ \epsilon_p(0) F_3 + P_3(0) \\ \epsilon_0\ \epsilon_p(0) F_3 + P_3(0) = \epsilon_0\ \epsilon_p(y) F_4 + P_4(y). \end{cases} \tag{4.26}$$

The electric fields are given as a solution to Eq. (4.26):

$$F_2 = \frac{1}{W\epsilon_0}[\epsilon_p(0)d_4(P(y) - P(x)) + \epsilon_p(y)d_3(P(0) - P(x)) - \epsilon_p(y)\epsilon_p(0)\epsilon_0 V],$$

$$F_3 = \frac{1}{W\epsilon_0}[\epsilon_p(y)d_2(P(x) - P(0)) + \epsilon_p(x)d_4(P(y) - P(0)) - \epsilon_p(x)\epsilon_p(y)\epsilon_0 V],$$

$$F_4 = \frac{1}{W\epsilon_0}[\epsilon_p(0)d_2(P(x) - P(y)) + \epsilon_p(x)d_3(P(0) - P(y)) - \epsilon_p(x)\epsilon_p(0)\epsilon_0 V],$$

$$W = d_2\epsilon_p(0)\epsilon_p(y) + d_3\epsilon_p(x)\epsilon_p(y) + d_4\epsilon_p(x)\epsilon_p(0). \tag{4.27}$$

Electric fields illustrated in Fig. 4.2 were calculated with Eq. (4.27) for the double barrier structure (30 Å) $Al_xGa_{1-x}N$/(40 Å) GaN/(30 Å) $Al_yGa_{1-y}N$.

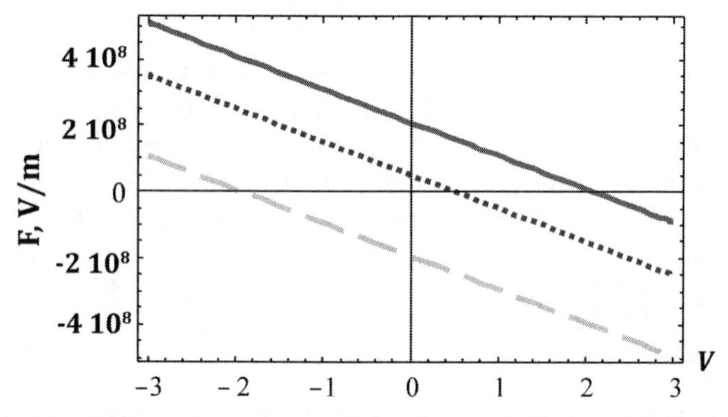

Figure 4.2 Voltage dependence of the electric fields; $x = 0.45$, $y = 0.3$. Solid line-barrier (region 2), dotted-barrier (region 4), dashed- well (region 3).

In an unbiased structure, the sign of an electric field in the well is opposite to those in the barriers. When applied voltage is large enough ($|V| > 2V$), the conduction band profile gets distorted so that all electric fields become of the same sign.

4.1.5 Spin Polarization

The numerical calculations have been done assuming the Dresselhaus coupling constant $\beta_l = 2 \times 10^{-31}$ eVm^3 [12] and Fermi energy $E_f = 0.1$ eV. The bulk k-linear spin-orbit constant λ and the difference between Rashba coefficients β_R-β_L can be neglected as their role in tunneling is much less than that of the

Dresselhaus term (4.9) that renormalizes the masses of tunneling electrons. Figure 4.3 illustrates the total vertical charge (J/q), Eq. (4.4), and spin J_S, Eq. (4.7), fluxes across the 30 Å/20 Å/30 Å double barrier $Al_{1-x}Ga_xN/GaN/Al_{1-y}Ga_yN$ structure, $F_L = 2.7 \times 10^5$ V/cm, $T = 300$ K, $\tau = 10^{-13}$ s.

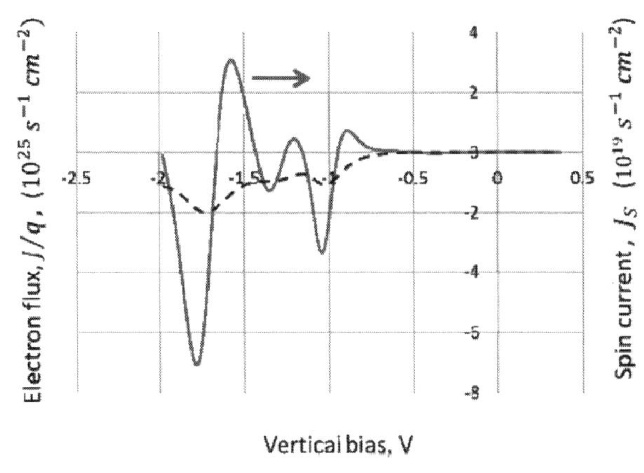

Figure 4.3 Current–voltage characteristics (dashed line) and spin current.

In this example, built-in electric fields make barrier shapes triangular (Zener tunneling) and increase transparency and spin flux as compared to the flat-band structure where an electrical polarization is not taken into account. Piezoelectric fields can be engineered by varying of Al-content in the barriers. This allows the spin injection efficiency in a spintronic device to be manipulated. Lattice mismatch induced electrical polarization could be responsible for the strain-dependent spin polarization that has been experimentally observed in Ref. [13].

Sheet spin density (spin polarization) in the region 5 can be estimated as $J_S\tau_S$, where τ_S is the spin relaxation time. Assuming that spin relaxation time in GaN-based structures $\tau_S = 0.25$ ps [14], it follows from the data shown in Fig. 4.3 that 1.5×10^{10} cm^{-2} spins are oriented at resonance voltage of −1.7V and moderate lateral electric field that corresponds to 20 kA/cm^2 current density in GaN. Longer spin relaxation time is favorable to the effects of spin injection and from this perspective $\tau_S = 100$ ps in InGaN multiple quantum well, reported in Ref. [15], may single out GaN/InGaN QW as a structure of choice for spintronic applications.

4.2 Spin Filtering in a Single-Barrier Tunnel Contact

The tunneling mechanism of spin injection does not necessarily rely on resonant transmission like described in previous section. Any spin-selective transmission process can serve that cause. In this section we consider spin selectivity that stems from over-barrier reflection in a single barrier contact.

Electron elastic scattering at the interface affects the current–voltage characteristics of the contacts. Since we are dealing with Rashba electron gas the potential scattering against non-magnetic impurities includes its spin-orbit part that might influence the discrimination between transmission amplitudes in two spin channels. In this section we discuss spin-polarized tunnel transmission through a biased GaN/AlGaN/GaN barrier. Both mechanisms of electron spin splitting in the barrier, Dresselhaus and Rashba, as well as the interface elastic scattering on non-magnetic impurities are taken into account.

4.2.1 Hamiltonian

We start with the structure that comprises single AlGaN barrier sandwiched between two thick GaN layers. The Hamiltonian (4.8) that accounts for interface scattering terms is given below:

$$H = \frac{h^2 k^2}{2m_{||}} + k_z\left[\frac{h^2}{2m_z} + \lambda_l(\sigma_x k_y - \sigma_y k_x)\right]k_z + W_0(\mathbf{r}_{||})\delta(z) + V(z)$$
$$+\left[W_1(\mathbf{r}_{||})\delta(z) + \left(\lambda + \frac{1}{2}\frac{\partial\beta}{\partial z} + \frac{\beta}{2}\frac{\partial}{\partial z}\right)\right](\sigma_x k_y - \sigma_y k_x), \quad (4.28)$$

The microscopic origin of the Rashba was considered in Chapter 2:

$$\beta(z) = \frac{P_1 P_2 \Delta_3}{(E_g + 2\Delta_2 - V(z))(E_g + \Delta_1 + \Delta_2 - V(z)) - 2\Delta_3^2},$$
$$\mathbf{r}_{||} = (x, y). \quad (4.29)$$

Matrix elements W_0 and W_1 describe spin-independent and spin-orbit parts of the impurity scattering amplitude, respectively. The impurity potential $W_{imp}(\mathbf{r})$ is an addition to the lattice periodic potential. Thus its matrix elements are similar to those generated by the periodic potential in the effective Hamiltonian, Eq. (4.28). The Hamiltonian contains the matrix elements

$$W_1(\mathbf{r}_{||}) = P_1 P_2 \frac{\partial}{\partial z} \frac{\langle Y|W_{\text{imp}}^{\text{s.o.}}(\mathbf{r})|Z\rangle}{(E_g + 2\Delta_2 - V(z))(E_g + \Delta_1 + \Delta_2 - V(z)) - 2\Delta_3^2},$$

$$W_0(\mathbf{r}_{||}) = \langle S|W_{\text{imp}}(\mathbf{r})|S\rangle,\ W_{\text{imp}}(\mathbf{r}) = \sum_i w(\mathbf{r} - \mathbf{R}_i), \tag{4.30}$$

where $|Z\rangle$, $|Y\rangle$, $|S\rangle$ are atomic basis wave functions. Integration in matrix elements (4.30) runs over the unit cell volume that contains an impurity, and $w(\mathbf{r})$ is the potential of a single impurity located at point $\mathbf{R}_i$, and $\Delta_{1,2,3}$ are the crystal field and spin-orbit parameters in the bulk material, respectively. Amplitudes W_0 and W_1 vary slowly with $\mathbf{r}_{||}$ on the scale of the average distance between impurities and can be considered to be random variables.

The conduction band profile is distorted by an external bias V_{ext} as shown in Fig. 4.4:

$$V(z) = \begin{cases} 0 & \text{left GaN } (l) \\ \Delta E_c - eFz, & \text{AlGaN barrier } (b) \\ -eV_{\text{ext}}, & \text{right GaN } (r) \end{cases}$$

$$F = V_{\text{ext}}/l, \tag{4.31}$$

ΔE_c is the conduction band offset, l is the barrier width.

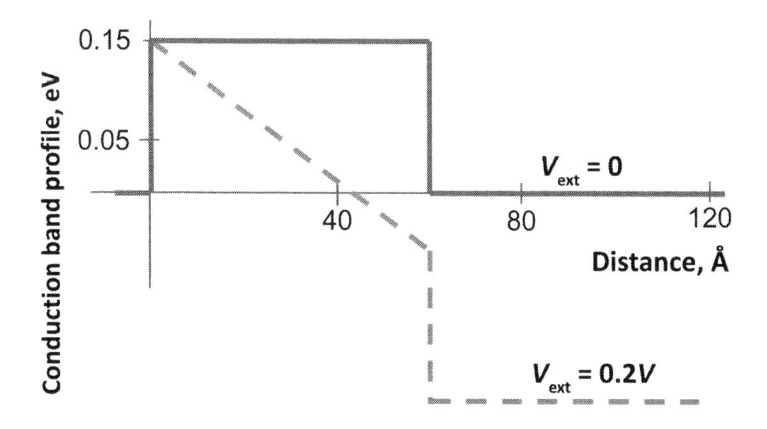

Figure 4.4 Conduction band profile in $Al_{0.2}Ga_{0.2}N$.

Unitary rotation to a new spin axis in the x–y plane transforms the eigenspinors and diagonalizes the Hamiltonian:

$$\Phi_{\pm}(\mathbf{r}_{||}, z) = \frac{1}{\sqrt{2}}\chi(z)u_{\pm}e^{i\mathbf{k}_{||}\mathbf{r}_{||}};\ u_{\pm} = \begin{pmatrix} \mp ie^{-i\varphi} \\ 1 \end{pmatrix};$$
$$H_{\pm} = \frac{\hbar^2 k^2}{2m_{||}} + k_z\left[\frac{\hbar^2}{2m_z^{\pm}}\right]k_z \mp \left(\lambda + \frac{1}{2}\frac{\partial\beta}{\partial z} + \frac{\beta}{2}\frac{\partial}{\partial z}\right)k + W_0(\mathbf{r}_{||})\delta(z) \mp W_1(\mathbf{r}_{||})\delta(z)k + V(z),$$
$$m_z^{\pm} = m_z(1 \mp 2\lambda_l m_z k_{||}/\hbar^2)^{-1}. \quad (4.32)$$

Since we consider scattering on non-magnetic impurities, there are no spin-flip processes in the new basis, i.e., the spin-orbit scattering amplitude W_1 does not mix the spin states $\Phi_{\pm}$.

Electron wave functions in layers, numbered in Eq. (4.31), can be expressed as follows:

$$\chi_{l\pm} = e^{ik_l z} + r_{\pm}e^{-ik_l z},\ \chi_{b\pm} = A_2 Ai(\rho_{\pm}) + B_2 Bi(\rho_{\pm}),$$
$$\chi_{r\pm} = t_{\pm}\ e^{ik_r(z-l)}, \quad (4.33)$$

where $A_i(\rho)$, $B_i(\rho)$ are the Airy functions, $r_{\pm}$, $t_{\pm}$ are the reflection and transmission amplitudes, respectively,

$$\hbar k_l = \sqrt{2m_z\left(E - \frac{\hbar^2 k^2}{2m_{||}}\right)};\ \hbar k_r = \sqrt{2m_z\left(E - \frac{\hbar^2 k^2}{2m_{||}} + eV_{\text{ext}}\right)};$$
$$\varrho_{\pm} = \left(\frac{2m_z^{\pm}}{\hbar^2 e^2 F^2}\right)^{\frac{1}{3}}(\Delta E_c - eFz - E). \quad (4.34)$$

4.2.2 Boundary Conditions and Spin-Selective Tunnel Transmission

In order to derive the boundary conditions that account for the random field pinned to an interface one needs to integrate the Schrödinger equation across the interface $z = 0$:

$$\int_{0_-}^{0_+} H_{\pm}\Phi_{\pm}dz = 0. \quad (4.35)$$

Multiplying (4.35) by the conjugate $\Phi_{\pm}^{*}$ and averaging over the interface area A, $(1/A)\int\ldots \exp[-(i\mathbf{k} + i\mathbf{k}')\mathbf{r}_{||}/2]d\mathbf{r}_{||}$, one arrives at the full set of matching conditions for wave functions and fluxes. Averaging the matching wave functions over the interface results in equal in-plane momenta on both sides of the interface $\mathbf{k} = \mathbf{k}'$. The boundary conditions at interface follow

$$\chi_l(0) = \chi_b(0),$$

$$\frac{1}{k_{Fl}}\frac{\partial \chi_l}{\partial z}\bigg|_{z=0} - \frac{m_l}{2m_r^{\pm}k_{Fl}}\frac{\partial \chi_b}{\partial z}\bigg|_{z=0} + \chi_l(0)[Z_0 \mp q(R + Z_1)] = 0,$$

$$q = k/k_{Fl},\ Z_0 = 2m_l V_0/(\hbar^2 k_{Fl}),\ Z_1 = 2m_1 V_1/\hbar^2,\ R = m_1\beta/\hbar^2. \quad (4.36)$$

Here the potential and spin-orbit scattering parameters V_0 and V_1 are zero Fourier components of random fields $W_{0,1}$: $V_{1,0} = (1/A)\int W_{1,0}(\mathbf{r}_{||})d\mathbf{r}_{||}$.

The dimensionless parameter Z_0 quantifies electron reflection at the interface and it has been used in the theory of contact phenomena [16]. One more parameter Z_1 naturally appears here to satisfy matching conditions in the presence of the spin-orbit impurity scattering that originates from the Rashba and Dresselhaus terms in electron spectrum [17]. It is assumed here that the random variables $Z_{0,1}$ have Gaussian distribution with the root mean square of $\Delta_{0,1} = 0.1 \langle Z_{0,1}\rangle$, where $\langle\cdots\rangle$ is the average over the random field distribution.

The tunnel transparency of the structure can be found using the transfer matrix method described in Section 4.1. The transfer matrix is composed of a set of four boundary conditions Eq. (4.36) for interfaces at $z = 0$ and $z = a$ (see Fig. 4.4).

Example numerical calculations have been performed for a 60 Å wide barrier with 20% Al content assuming Fermi energy of $E_F = 85$ meV. Bulk k-linear spin-orbit λ has been neglected as its role in tunneling is much less important than that of the Dresselhaus and Rashba SIA terms. The parameter $R \approx 10^{-3}$ has been estimated from β Eq. (4.29). One can obtain an order of magnitude estimation of parameters $Z_{0,1}$ assuming the Coulomb impurity potential energy scale of $C_0 \approx 1$ eV and spin-orbit characteristic energy of $C_1 \approx 1$ meV. Estimations of the scattering parameters (4.36) gives $V_0/a \approx C_0$ and $V_1 k_{Fl}/a \approx C_1$, $Z_0 \approx 3$, $Z_1 \approx 3 \times 10^{-3}$, where a is the lattice constant. Figure 4.5 illustrates the averaged transmission $\langle t_+\rangle$ calculated for the structure shown in Fig. 4.4.

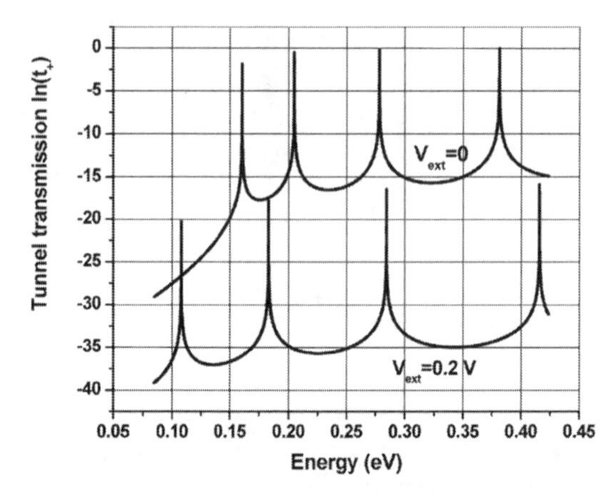

Figure 4.5 Tunnel transmission as a function of energy, $k = 0.3k_{Fl}$, $<Z_0> = 3$; $<Z_1> = 0.003$.

Over-barrier reflection is at the origin of resonances in the transmission coefficient, making the two barrier boundaries act as a semi-transparent Fabry–Perot resonator with the quality factor dependent on the reflection coefficients on both interfaces. The first resonance in an unbiased structure occurs at $E = 0.16$ eV, just above the top of the barrier. Finite bias makes the shape of the barrier triangular and then shifts the resonance to lower energies. Single-mode tunnel spin polarization is defined as $P = \dfrac{<t_+> - <t_->}{<t_+> + <t_->}$ and is shown in Fig. 4.6 as a function of the applied voltage.

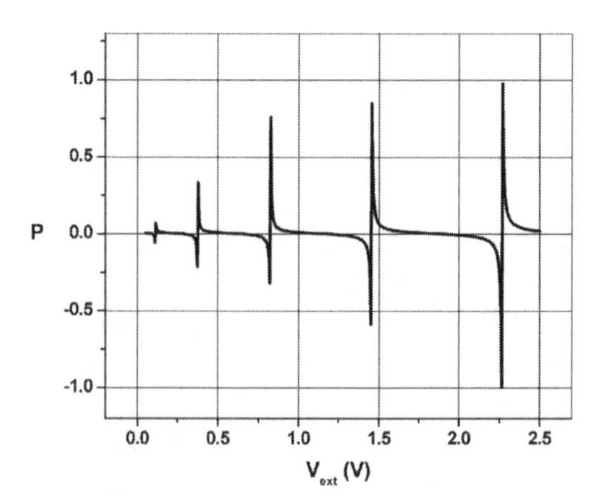

Figure 4.6 Spin polarization vs. applied voltage, $E = 1.5E_F$, $k = 1.1k_{Fl}$, $<Z_0> = 3$; $<Z_1> = 0.003$.

Resonances in transmission (Fig. 4.5) and spin polarization (Fig. 4.6) originate from over-barrier electron reflection and look similar to those that occur in double-barrier structures, where they stem from resonance tunneling.

Examples of single-mode spin polarization dependent on scattering parameters $<Z_0>$ and $<Z_1>$ are illustrated in Figs. 4.7 and 4.8, respectively.

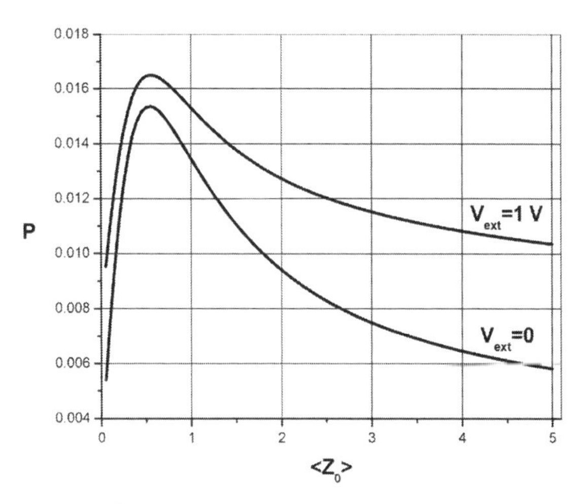

Figure 4.7 Dependence of spin polarization on a potential scattering parameter, $E = 1.5\ E_F$, $k = 1.1\ k_{Fl}$, $<Z_1> = 0.003$.

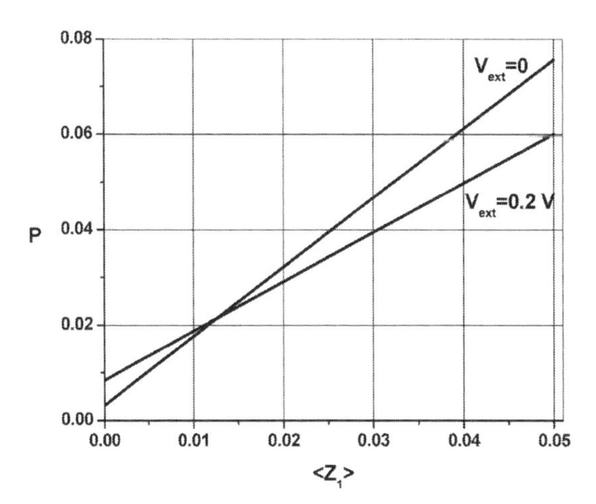

Figure 4.8 Dependence of spin polarization on a spin-orbit scattering parameter, $E = 1.5\ E_F$, $k = 1.1\ k_{Fl}$, $<Z_0> = 3$.

As shown in Fig. 4.8, the spin-orbit scattering parameter Z_1 affects the spin polarization.

The single-mode spin polarization by itself does not mean that spin injection occurs when the tunnel current flows across the barrier: In equilibrium, the summation over all in-plane wave vectors makes the total spin polarization zero. Spin injection occurs if some non-equilibrium process violates in-plane symmetry (for instance, in-plane electric current in the region (r) or (l)). The resulting in-plane spin current is perpendicular to a lateral electric field F_L (see Section 4.1.2).

The magnitude of spin current, produced by tunneling electrons of energy E, can be written as

$$J_S(x) = \frac{q\tau F_L k_{Fl}^3}{32\pi^2 m_1} \int_0^{Q_{max}(x)} q^2 dq \frac{\partial f_l}{\partial x} (<t_+> - <t_->), \tag{4.37}$$

where $x = E/E_F$, f_l is the equilibrium electron distribution function in the region (l). The upper limit $Q_{max}(x)$ is determined by the region where both incident and outgoing electron wave vectors are real.

Assuming a momentum relaxation time of $\tau = 10^{-13}$ s and the lateral field of $F_L = 10^4$ V/cm the spin current calculated with Eq. (4.37) is shown in Fig. 4.9.

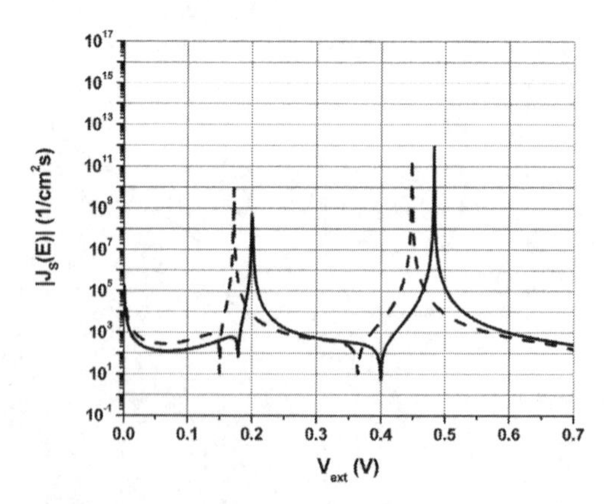

Figure 4.9 Spin flux as a function of applied voltage. $<Z_0> = 3$, $<Z_1> = 0.003$ solid line: $E = 0.9E_F$, dashed: $E = E_F$.

In conclusion, even single barrier structures may deliver resonances in spin filtering due to over-barrier electron reflection.

The spin-orbit part of interface impurity scattering plays an important role in spin-selective electron transmission and may enhance spin polarization efficiency. This makes single-barrier ballistic contacts competitive with double-barrier structures where resonances stem from resonance tunneling. This is especially important in spintronic applications of wide bandgap semiconductors, where high-quality double-barrier resonant tunneling diodes have yet to be reported (see review Ref. [18]).

Problems

4.1 Find energy levels and wave functions of an electron in triangular quantum well.

4.2 Show that resonance tunneling transparency in a double-barrier structure reaches its maximum at $T_1 = T_2$, where $T_{1,2}$ are transmission coefficients of barriers 1 and 2, respectively.

References

1. Schmidt, G, Ferrand, D, Molenkamp, LW, Filip, AT, van Wees BJ (2000) Fundamental obstacle for electrical spin injection from a ferromagnetic metal into a diffusive semiconductor, *Phys Rev B*, **62,** R4790.
2. Rashba EI (2000) Theory of electrical spin injection: Tunnel contacts as a solution of the conductivity mismatch problem, *Phys Rev B*, **62**, R16267.
3. Rashba EI (2004) Spin currents, spin populations, and dielectric function of noncentrosymmetric semiconductors, *Phys Rev B*, **70**, 161201(R).
4. Ivchenko EL, Pikus GE (1978) New photogalvanic effect in gyrotropic crystals, *JETP Lett*, **27**, 604–607.
5. Magarill LI, Entin MV (2000) Spin orientation of two-dimensional electrons in electric field, *JETP Lett*, **72**(3), 195–200.
6. Kalevich VK, Korenev VL (1990) Effect of electric current on the optical orientation of 2D electrons, *JETP Lett*, **52**, 230–235.
7. Alekseev PS, Chistyakov VM, Jassievich IN (2006) Electric-field effect on the spin-dependent resonance tunneling, *Semiconductors*, **40**(12), 1402–1408 (*Fiz Tekhn Poluprovodnikov*, **40**, 12, 1436–1442).
8. Litvinov VI, Manasson A, Pavlidis D (2004) Short-period intrinsic Stark GaN/AlGaN superlattice as a Bloch oscillator, *Appl Phys Lett*, **85**, 600–602.

9. Wei H, Zunger A (1998) Calculated natural band offsets of all II–VI and III–V semiconductors: Chemical trends and the role of cation *d*-orbitals, *Appl Phys Lett*, **72**, 2011–2013.
10. Kumagai M, Chuang SL, Ando H (1998) Analytical solutions of the block-diagonalized Hamiltonian for strained wurtzite semiconductors, *Phys Rev B*, **57**, 15303.
11. Ambacher O, Majewski J, Miskys C, et al. (2002) Pyroelectric properties of Al(In)GaN/GaN hetero- and quantum well structures, *J Phys Condens Mater*, **14**, 3399–3434.
12. Cheng H, Biyikli N, Özgür Ü, Kurdak Ç, Morkoç H, Litvinov VI (2008) Measurement of linear and cubic spin–orbit coupling parameters in AlGaN/AlN/GaN heterostructures with a polarization-induced two-dimensional electron gas, *Physica E*, **40**, 1586–1589.
13. Chang HJ, Chen TW, Chen JW, et al. (2007) Current and strain-induced spin polarization in InGaN/GaN superlattices, *Phys Rev Lett*, **98**, 136403.
14. Kuroda T, Yabushita T, Kosuge T, et al. (2004) Subpicosecond exciton spin relaxation in GaN, *Appl Phys Lett*, **85**, 3116–3118.
15. M. Jullier, A. Vinatieri, M. Colloci, et al (1999) Slow spin relaxation observed in InGaN/GaN multiple quantum wells, *Phys State Solidi* (*b*) **216**, 341–345.
16. Hu CM, Matsuyama T (2001) Spin injection across a heterojunction: A ballistic picture, *Phys Rev Lett*, **87**, 066803.
17. Litvinov VI (2010), Resonance spin filtering due to overbarrier reflection in a single barrier contact, *Phys Rev B*, **82**, 109035.
18. Litvinov VI (2010) Resonant tunneling in III-Nitrides, *Proc IEEE*, **98**, 1249–1254.

Chapter 5

Exchange Interaction in Semiconductors and Metals

Material templates used in spintronics include various metallic magnetic materials that show a direct magnetoresistive effect or metallic junctions with tunneling magnetoresistance. Also, nonmagnetic semiconductors are recognized as a material template for spintronic devices since the spin-polarized electrons injected from metallic ferromagnetic contact may have a long spin lifetime enabling spin transfer to another ferromagnetic electrode where the spin information can be read out. The conductivity mismatch between a ferromagnetic metal contact and a semiconductor active region degrades the efficiency of spin injection. There are two basic ways to avoid the conductivity mismatch: One is to use tunnel contacts, another is to replace ferromagnetic metals with semiconductor alloys containing transition-metal atoms such as Fe, Mn, and Co providing they show room-temperature ferromagnetism [1,2]. Ferromagnetic semiconductors would benefit device performance as the all-semiconductor structure can be grown lattice matched to the other part of the device, reducing the defect density, and increasing the lifetime of non-equilibrium carriers and the spin relaxation time.

Ferromagnetism in diluted magnetic semiconductors is the subject of intense theoretical and experimental studies with respect

Wide Bandgap Semiconductor Spintronics
Vladimir Litvinov

ISBN 978-981-4669-70-2 (Hardcover), 978-981-4669-71-9 (eBook)
www.panstanford.com

to their various applications in spintronics. The central point of the study is the nature of magnetic interaction between electrons and localized spins and also between localized magnetic moments. In this chapter, we discuss several mechanisms of indirect exchange interaction between magnetic atoms in metals and semiconductors. The details of this interaction are important as they determine the ferromagnetic critical temperature and help in the engineering of high-temperature ferromagnetic semiconductors.

5.1 Direct Exchange Interaction

A Hamiltonian of two electrons coupled by Coulomb interaction has the form

$$H = H_1(\mathbf{r}_1) + H_2(\mathbf{r}_2) + U(\mathbf{r}_1, \mathbf{r}_2), \tag{5.1}$$

where $H_{1,2}$ are one-particle energy operators and $U(\mathbf{r}_1, \mathbf{r}_2)$ is Coulomb interaction. Basis wave functions are eigenfunctions of one-particle parts of the Hamiltonian: $\varphi_1(\mathbf{r})$ and $\varphi_2(\mathbf{r})$. The wave function for the pair of non-interacting electrons $\Psi(\mathbf{r}_1, \mathbf{r}_2)$ can be expressed through products of one-electron functions $\varphi_1(\mathbf{r})$ and $\varphi_2(\mathbf{r})$. Composing the products one has to account for electron spin $\left(s_1 = s_2 = \frac{1}{2}\right)$ and write two-electron wave functions corresponding to total spins $S = 0$ (antiparallel spins in the pair) and $S = 1$ (parallel spins). Swapping the coordinates of electrons 1 and 2 changes the wave function by a factor $(-1)^S$: $\Psi(\mathbf{r}_1, \mathbf{r}_2) = (-1)^S\ \Psi(\mathbf{r}_2, \mathbf{r}_1)$, so, the wave function should be symmetric for $S = 0$ and anti-symmetric for $S = 1$:

$$\Psi_S(\mathbf{r}_1, \mathbf{r}_2) = \varphi_1(\mathbf{r}_1)\,\varphi_2(\mathbf{r}_2) \pm \varphi_1(\mathbf{r}_2)\,\varphi_2(\mathbf{r}_1) \tag{5.2}$$

The total energy of the two electrons is given as

$$E = E_1 + E_2 + \Delta E_S, \tag{5.3}$$

where $E_{1,2}$ are one-particle energies of non-interacting electrons (eigenvalues of $H_{1,2}$, respectively), and ΔE_S is the first-order perturbation correction induced by Coulomb interaction:

$$\Delta E_S = \int \Psi_S^*(\mathbf{r}_1, \mathbf{r}_2) U(\mathbf{r}_1, \mathbf{r}_2)\, \Psi_S(\mathbf{r}_1, \mathbf{r}_2)\, d\mathbf{r}_1 d\mathbf{r}_2 \tag{5.4}$$

Making use of Eq. (5.2) to calculate ΔE_S one obtains $\Delta E_0 = C + J_{ex}$, $\Delta E_1 = C - J_{ex}$ where the direct Coulomb shift C and the exchange correction J_{ex} are given below:

$$C = \int |\varphi_1(\mathbf{r}_1)|^2 |\varphi_2(\mathbf{r}_2)|^2 U(\mathbf{r}_1, \mathbf{r}_2) d\mathbf{r}_1 d\mathbf{r}_2$$
$$J_{ex} = \int \varphi_1^*(\mathbf{r}_2)\varphi_2^*(\mathbf{r}_1) U(\mathbf{r}_1, \mathbf{r}_2)\varphi_1(\mathbf{r}_1)\varphi_2(\mathbf{r}_2) d\mathbf{r}_1 d\mathbf{r}_2. \tag{5.5}$$

Exchange corrections $\pm J_{ex}$ depend on the total spin of the electrons. The total spin vector $\mathbf{S} = \mathbf{s}_1 + \mathbf{s}_2$ squared gives the relation

$$\mathbf{s}_1\mathbf{s}_2 = \frac{1}{2}(\mathbf{S}^2 - \mathbf{s}_1^2 - \mathbf{s}_2^2). \tag{5.6}$$

Replacing the square of the momenta by its eigenvalue in Eq. (5.6), we have eigenvalues of the product

$$(\mathbf{s}_1\mathbf{s}_2)_{ev} = \frac{1}{2}[S(S+1) - s_1(s_1+1) - s_2(s_2+1)] = \begin{cases} -\frac{3}{4}, S = 0 \\ \frac{1}{4}, S = 1 \end{cases}. \tag{5.7}$$

Exchange energy corrections can be expressed with the help of (5.7) as

$$\pm J_{ex} = -J_{ex}\left(\frac{1}{2} + 2(\mathbf{s}_1\mathbf{s}_2)_{ev}\right). \tag{5.8}$$

These are eigenvalues of the Hamiltonian

$$H_{ex} = -J_{ex}\left(\frac{1}{2} + 2\mathbf{s}_1\mathbf{s}_2\right). \tag{5.9}$$

The Hamiltonian (5.9) was obtained by Dirac and then its spin-dependent part was used by Heisenberg in the theory of ferromagnetism in solids. In magnetic materials the Heisenberg model $H = -J_{ex}\mathbf{s}_1\mathbf{s}_2$ describes the direct exchange interaction between localized spins $\mathbf{s}_1$ and $\mathbf{s}_2$ and favors the parallel spin orientation if $J_{ex} > 0$. Exchange interaction between atoms takes place only if their electron wave functions overlap, so direct Heisenberg interaction is adequate for magnetic materials where the spin moments of the host atoms are placed at a distance of the order of the lattice constant. This is not the case in magnetically doped metals, semiconductors, or magnetic alloys where the content of magnetic atoms is a small fraction of that of the host atoms. Still, the coupling between distant spins exists as it is mediated by the host: free electrons in metals and degenerate semiconductors, or nonmagnetic atoms in dielectrics. This type of coupling is called an indirect exchange interaction.

In this chapter, we discuss some features of indirect exchange in metals and III-V narrow and wide bandgap semiconductors as well as the picture of ferromagnetic phase transition in magnetically doped semiconductors.

5.2 Indirect Exchange Interaction

In metals and semiconductors, indirect exchange appears as a result of coupling between an impurity spin and a free *s*-electron spin. The localized spin originates from the unfilled *d*-shell of a transition metal atom or the *f*-shell of a rare earth element, so the coupling is referred as *s–d* or *s–f* interaction. As the inner unfilled orbitals are more localized than the valence electron *s*-states, the *s–d* interaction occurs in close proximity to a magnetic impurity and can be well approximated by a contact interaction. The full one-electron Hamiltonian is given as

$$H(\mathbf{r}) = H_e(\mathbf{r}) + H_{sd}(\mathbf{r}),$$
$$H_e(\mathbf{r}) = \frac{p^2}{2m_0} + V(\mathbf{r}), H_{sd}(\mathbf{r}) = -\frac{J}{n}\sum_i \sigma\mathbf{S}_i\delta(\mathbf{r}-\mathbf{R}_i), \tag{5.10}$$

where $\mathbf{p} = -i\hbar\partial/\partial\mathbf{r}$, $\sigma = \{\sigma_x, \sigma_y, \sigma_z\}$ are the Pauli matrices, m_0 is the free electron mass, $V(\mathbf{r})$ is the electron energy in the lattice periodic potential, J is the coupling constant of an isotropic *s–d* interaction, $n = N/V$ is the density of the host atoms, and $\mathbf{R}_i$ is the impurity position in the lattice. In the effective mass approximation, the free electron Hamiltonian can be rewritten as $H_e(\mathbf{r}) = p^2/2m$, where the effective mass m includes effects of the periodic field $V(\mathbf{r})$ and gives the correct energy dispersion in the vicinity of the edge of the conduction band. This approximation works well in metals and degenerate semiconductors if energy corrections from remote energy bands are negligible.

Hamiltonian (5.10) is the 2 × 2 matrix in an electron spin space, so H_e is understood as to be multiplied by the unit matrix $I = \mathrm{diag}(1,1)$.

In what follows, we consider the two-band model for electrons that allows us to derive the indirect exchange in both metals and semiconductors. The electron wave function can be expressed as an expansion over the Kohn–Luttinger basis:

$$\Psi(\mathbf{r}) = \frac{1}{\sqrt{V}}\sum_{\mathbf{k}} a_{\mathbf{k}} e^{i\mathbf{kr}}, \; a_{\mathbf{k}} = \begin{pmatrix} u_1(\mathbf{r})C_{\mathbf{k}1\uparrow} \\ u_1(\mathbf{r})C_{\mathbf{k}1\downarrow} \\ u_2(\mathbf{r})C_{\mathbf{k}2\uparrow} \\ u_2(\mathbf{r})C_{\mathbf{k}2\downarrow} \end{pmatrix}, \tag{5.11}$$

where the column $a_{\mathbf{k}}$ describes up and down electron spin states in the bands $j = 1,2$, $u_j(\mathbf{r})$ is the Bloch amplitude at the edge of the band j.

In the spirit of the second quantization representation, we consider coefficients $C_{\mathbf{k}js}(C^{+}_{\mathbf{k}js})$ as Fermi annihilation (creation) operators acting in the occupation number space, destroying (creating) an electron with the quantum numbers $\mathbf{k}$, j, s. In this representation, the wave function $\Psi(\mathbf{r})$ becomes a field operator. In the representation (5.11) the Hamiltonian becomes

$$H = \int d\mathbf{r}\, \Psi^{+}(\mathbf{r})H(\mathbf{r})\Psi(\mathbf{r}) = H_e + H_{sd},$$

$$H_e = \sum_{\mathbf{k}} C^{+}_{\mathbf{k}}[\hat{E}(\mathbf{k})I]C_{\mathbf{k}},\; H_{sd} = \frac{1}{N}\sum_{\mathbf{kq}}\sum_{i} \exp(i\mathbf{q}\mathbf{R}_i)C^{+}_{\mathbf{k}}[\hat{J}\sigma\mathbf{S}_i]C_{\mathbf{k}+\mathbf{q}},$$

$$C_{\mathbf{k}} = \begin{pmatrix} C_{\mathbf{k}1\uparrow} \\ C_{\mathbf{k}1\downarrow} \\ C_{\mathbf{k}2\uparrow} \\ C_{\mathbf{k}2\downarrow} \end{pmatrix}, \tag{5.12}$$

where $\hat{E}_{jl}(\mathbf{k})$ and $\hat{J}_{jl} = Ju^{*}_{j}(\mathbf{R}_i)u_l(\mathbf{R}_i)$ are the matrices operating in the space of two semiconductor bands (j, l = 1, 2). The unit matrix I and spin matrices σ act in the 2 × 2 spin space. Since the matrices in Eq. (5.12) act in different spaces, the matrix products $\hat{E}(\mathbf{k})I$ and $\hat{J}\boldsymbol{\sigma}$ are understood to be direct (Kronecker) products. Here $\hat{E}(\mathbf{k})$

and $\hat{J}$ are the energy spectrum and s–d interaction constant, respectively.

In a single-band metal, $\hat{E}(\mathbf{k})$ and $\hat{J}$ are the scalars, $\boldsymbol{E}(\mathbf{k})=\dfrac{h^2\mathbf{k}^2}{2m}$, $\hat{J}=J$. In a two-band semiconductor, $\hat{J}$ and $\hat{E}(\mathbf{k})$ are 2 × 2 matrices:

$$\hat{E}(\mathbf{k})=\begin{pmatrix} E_1(\mathbf{k}) & E_{12} \\ E_{21} & E_2(\mathbf{k}) \end{pmatrix},\quad \hat{J}=\begin{pmatrix} J_1 & J_{12} \\ J_{12} & J_2 \end{pmatrix},$$

$$E_1(\mathbf{k})=\hbar^2\mathbf{k}^2/2m_c+E_g/2,$$

$$E_2(\mathbf{k})=-\hbar^2\mathbf{k}^2/2m_v-E_g/2, \tag{5.13}$$

where E_g is the bandgap, the reference energy is chosen in the middle of the bandgap.

In III-V semiconductors where k–p perturbation theory is being used to calculate the energy spectrum, the non-diagonal matrix elements $E_{12}=E_{21}=kP_{\text{cv}}$ make the energy dispersion non-parabolic (see Chapter 1). At small k the non-parabolic corrections just renormalize the effective masses of electrons and holes. So the off-diagonal term E_{12} will be neglected in further considerations.

The s–d coupling in (5.12) describes the electron scattering against an impurity spin located at $\mathbf{R}_i$; in this process both the electron and the impurity change their spin orientations. The indirect exchange interaction between a pair of localized spins appears as a second order energy correction with respect to H_{sd}. It is convenient to express the second order correction by the diagram shown in Fig. 5.1:

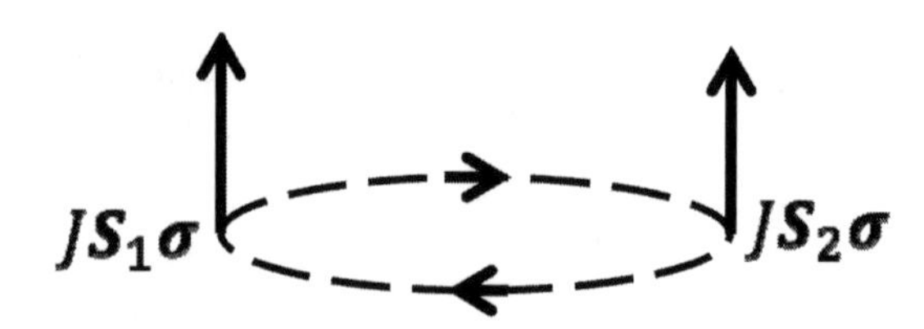

Figure 5.1 Indirect exchange interaction.

The diagram depicts the process in which an impurity spin creates a virtual electron–hole pair, which then propagates to another impurity and annihilates there. Two localized spins exchange the virtual electron–hole pair, maintaining the indirect exchange interaction between each other. The process is similar to the vacuum polarization process in which process a photon creates a virtual electron–positron pair. The process, shown in Fig. 5.1, can also be called vacuum polarization if a vacuum is understood to be a Fermi surface in a single band metal or fully occupied valence band and the empty conduction band in a semiconductor at a temperature of absolute zero.

The propagation of electrons and holes can be described by the Green function. In the finite temperature Matsubara technique, the Green function is expressed as [3]

$$G(\mathbf{k}, \omega_m) = \begin{pmatrix} \dfrac{I}{i\omega_m + \mu - E_1(\mathbf{k})} & 0 \\ 0 & \dfrac{I}{i\omega_m + \mu - E_2(\mathbf{k})} \end{pmatrix}, \tag{5.14}$$

where the unit matrix I acts in the spin space, $\omega_m = (2m + 1)\pi T$ is the Matsubara frequency, $m = 0, \pm 1, \ldots$, T is the temperature in energy units, μ is the chemical potential. Each arrow line in Fig. 5.1 corresponds to an electron Green function and each vertex contains the s–d coupling matrix $\hat{J}\mathbf{S}_{1,2}\boldsymbol{\sigma}\exp(i\mathbf{q}\mathbf{R}_{1,2})$. The analytical expression follows from the rules of diagrammatic technique:

$$H = \frac{T}{N^2} \sum_{m=-\infty}^{\infty} \sum_{\mathbf{k},\mathbf{q}} \exp(i\mathbf{q}\mathbf{R}) Tr \{\hat{J}(\boldsymbol{\sigma}\mathbf{S}_1) G(\mathbf{k}, \omega_m) \hat{J}(\boldsymbol{\sigma}\mathbf{S}_2) G(\mathbf{k} + \mathbf{q}, \omega_m)\}, \tag{5.15}$$

where $\mathbf{R} = \mathbf{R}_1 - \mathbf{R}_2$.

The expression, Eq. (5.15), is written in a general form where the dimension of the matrices depends on the type of electron spectrum in the host. The operation $Tr = Tr_s Tr_j$ implies summation over the diagonal matrix elements, that is summation over electron quantum numbers: bands (j) and spins (s). The operation Tr_j gives

$$\begin{aligned} H = \frac{T}{N^2} \sum_{m=-\infty}^{\infty} \sum_{k,q} \exp(i\mathbf{q}\mathbf{R}) [J_1^2 G_{11}(k, \omega_m) G_{11}(\mathbf{k} + \mathbf{q}, \omega_m) \\ + J_2^2 G_{22}(\mathbf{k}, \omega_m)\, G_{22}(\mathbf{k} + \mathbf{q}, \omega_m) \\ + 2|J_{12}|^2 G_{11}(\mathbf{k}, \omega_m) G_{22}(\mathbf{k} + \mathbf{q}, \omega_m)]\ Tr_s\{(\sigma\mathbf{S}_1)(\sigma\mathbf{S}_2)\}, \end{aligned} \tag{5.16}$$

and, after summation over the spin indexes, $Tr_s\{(\boldsymbol{\sigma}\mathbf{S}_1)(\boldsymbol{\sigma}\mathbf{S}_2)\} = \mathbf{S}_1\mathbf{S}_2$, one gets the indirect exchange interaction in the form

$$H = V(R)\mathbf{S}_1\mathbf{S}_2,$$

$$V(R) = \frac{T}{N^2}\sum_{m=-\infty}^{\infty}\sum_{k,q}[J_1^2 G_{11}(\mathbf{k},\omega_m)G_{11}(\mathbf{k}+\mathbf{q},\omega_m)$$
$$+J_2^2 G_2(\mathbf{k},\omega_m)G_{22}(\mathbf{k}+\mathbf{q},\omega_m)$$
$$+2|J_{12}|^2 G_{11}(\mathbf{k},\omega_m)\,G_{22}(\mathbf{k}+\mathbf{q},\omega_m)]\exp(i\mathbf{qR}). \tag{5.17}$$

The calculation of the frequency sums can be done using the relation

$$T\sum_m g(i\omega_m) = -\sum_i Res[g(x_i)f(x_i)],$$

$$f(x) = \frac{1}{2}\tanh(x/2T), \tag{5.18}$$

where residues are taken in the poles of $g(x)$. At $T \to 0$ one may use frequency integration with the substitution

$$i\omega_m \to \omega + i\delta\, sgn(\omega),\, T\sum_m \to \frac{1}{2\pi i}\int_{-\infty}^{\infty} d\omega, \delta \to 0. \tag{5.19}$$

The frequency integrals in (5.17) contain terms

$$I_i = \int_\Gamma \frac{d\omega}{(\omega+\mu-E_i(\mathbf{k}))\,(\omega+\mu-E_i(\mathbf{k}+\mathbf{q})},$$

$$I_{12} = \int_\Gamma \frac{d\omega}{(\omega+\mu-E_1(\mathbf{k}))(\omega+\mu-E_2(\mathbf{k}+\mathbf{q})} \tag{5.20}$$

which can be calculated by closing the integration path in the upper or lower frequency half-plane. At this point, it is important to locate the positions of the poles. The actual situation depends on the energy spectrum in the host. Below we consider possible options.

5.3 Three-Dimensional Metal: RKKY Model

In a simple metal we deal with a single band, μ is the Fermi energy. The only relevant term in Eq. (5.17) is that proportional to I_1. If the poles of the integrand are both positive or negative, one can close the integration path so that no poles lie inside the closed path and then the integral is equal to zero. This means that the nonzero I_1 is possible only if $p_1 = \varepsilon(\mathbf{k}) - \mu < 0$ and $p_2 = \varepsilon(\mathbf{k}+\mathbf{q}) - \mu > 0$, or vice versa. This corresponds to virtual electron–hole pair excitation across the Fermi level.

Integral I_1 can be calculated by closing the integration path in the upper frequency half-plane as shown in Fig. 5.2:

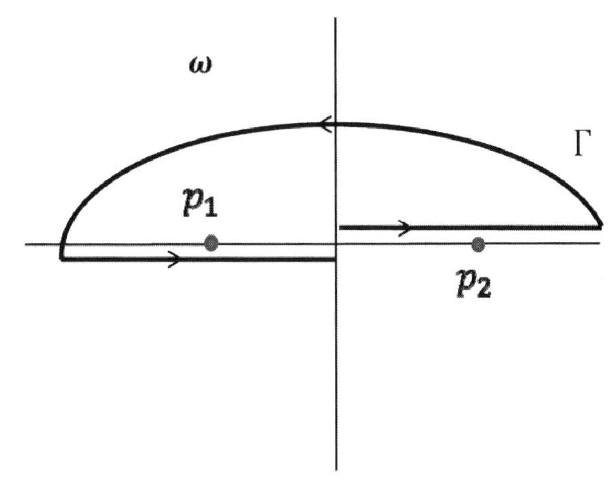

Figure 5.2 Integration path and poles in integral I_1.

$$I_1 = \frac{1}{2\pi i}\int_\Gamma \frac{\exp(i\omega\tau)d\omega}{(\omega + \mu - \varepsilon(\mathbf{k}))(\omega + \mu - \varepsilon(\mathbf{k}+\mathbf{q}))}\bigg|_{\tau\to 0}$$
$$= \frac{1}{\varepsilon(\mathbf{k}) - \varepsilon(\mathbf{k}+\mathbf{q})}, \varepsilon(\mathbf{k}) = \frac{\hbar^2\mathbf{k}^2}{2m} \tag{5.21}$$

It is instructive to note that the same result follows from the relation (5.18) in the limit $T \to 0$:

$$T\sum_m \frac{1}{(i\omega_m + \mu - \varepsilon(\mathbf{k})\,(i\omega_m + \mu - \varepsilon(\mathbf{k}+\mathbf{q}))} \tag{5.22}$$
$$= \frac{f(\varepsilon(\mathbf{k})-\mu) - f(\varepsilon(\mathbf{k}+\mathbf{q})-\mu)}{\varepsilon(\mathbf{k}) - \varepsilon(\mathbf{k}+\mathbf{q})} \to \frac{1}{\varepsilon(k) - \varepsilon(\mathbf{k}+\mathbf{q})}.$$

It follows from the Eq. (5.22) that at $T = 0$ the sum is equal to zero if both the initial and the final scattering energy, $\varepsilon(\mathbf{k})$ and $\varepsilon(\mathbf{k} + \mathbf{q})$, happen to be on the same side with respect to the Fermi level, that is equivalent to our previous notion of a zero integral if the poles are on the same side of the real axis in the frequency integral (5.21).

After integration over frequency the range function (5.17) is expressed as:

$$V_{\mathrm{RKKY}}(R)=\frac{J_1^2}{n^2(2\pi)^6}\int d\mathbf{k}d\mathbf{k}'\,\frac{\cos[\mathbf{R}(\mathbf{k}-\mathbf{k}')]}{\varepsilon(\mathbf{k})-\varepsilon(\mathbf{k}')},\tag{5.23}$$

Integration limits over the wave vectors are determined by the conditions under which the initial wave vectors lie inside the Fermi sphere, $k \le k_{\mathrm{F}}$, and the momentum transfer due to scattering on a localized spin, q, spans the interval $(0, \infty)$. The calculation of the integrals yelds the RKKY result [4]:

$$\begin{aligned}
V_{\mathrm{RKKY}}(R) \\
&=\frac{J_1^2}{n^2(2\pi)^6}\int_0^{2\pi}d\varphi d\varphi'\int\frac{k'^2k^2dkdk'}{\varepsilon(k)-\varepsilon(k')}\int_{-1}^{1}\cos(Rkx)\cos(Rk'x')dxdx' \\
&=\frac{mJ_1^2}{4\pi^4n^2\hbar^2R^2}\int_0^{K_{\mathrm{F}}}kdk\int_0^{\infty}\frac{\sin(Rk)\sin(Rk')}{k^2-k'^2}k'dk' \\
&=\frac{mJ_1^2}{8n^2\pi^3\hbar^2R^2}\int_0^{k_{\mathrm{F}}}\cos(Rk)\sin(Rk)kdk \\
&=\frac{mJ_1^2}{8n^2\pi^3\hbar^2R^2}\left[-\frac{k_{\mathrm{F}}\cos(2k_{\mathrm{F}}R)}{4R}+\frac{\sin(2k_{\mathrm{F}}R)}{8R^2}\right].
\end{aligned}\tag{5.24}$$

In Eq. (5.24) we used the relation

$$\int_0^{\infty}\frac{\sin(Rk')}{k^2-k'^2}k'dk'=\frac{\pi}{2}\cos(Rk).\tag{5.25}$$

As we assumed a parabolic energy spectrum that is valid for small wave vectors $k << 1/a$, a being the lattice constant, one can use the range function (5.24) in the range $R >> \alpha$. At large distances, $Rk_{\mathrm{F}} >> 1$, $V_{\mathrm{RKKY}}(R)$ decays as $1/R^3$ and oscillaltes with the spatial period π/k_{F}. If the carrier density (electrons or holes) n_c is low, we deal with the short-distance regime, $RK_{\mathrm{F}} << 1$. The interaction is negative, so it favors the ferromagnetic ordering and scales as $k_{\mathrm{F}} \sim n_c^{1/3}$.

In an impure metal, electron scattering can be taken into account introducing the imaginary term in Eq. (5.23): $\omega \to \omega + i/\tau$, τ is the momentum relaxation time with respect to scattering on non-magnetic centers. As a result, the RKKY range function acquires an additional decay factor [5,6]:

$$\widetilde{V}(R) = V_{\mathrm{RKKY}}(R)\exp(-R/l), \tag{5.26}$$

where $l = \hbar k_{\mathrm{F}}\tau/m$ is the electron mean free path.

5.4 RKKY Interaction in One and Two Dimensions

Usually, the frequency summation is done first and then the double integral over momenta remains to be calculated. This sequence of steps was used in Sections 5.2 and 5.3 in order to get three dimensional (3D) RKKY. In one-dimensional (1D) metals the integrand in Eq. (5.23) contains a strong singularity when wave vectors tend to zero. The non-analytical behavior of the integrand makes the result dependent on the order of integration over the electron momentum k and the momentum transfer $q = k - k'$. That is the subject of detailed discussion in Ref. [7].

An alternative method of calculation suggests integrating over momenta as the first step and then performing the frequency summation. This method has been applied to bulk narrow-gap semiconductors and quantum wells, with no carriers at $T = 0$, presenting Bloembergen–Rowland-type of interactions [8] and to RKKY interaction in metals of various dimensions in Refs. [9, 10]. As long as one deals with 2D or 3D electron system, it does not matter what sequence of calculations is chosen in Eq. (5.23). However, in a 1D metal the sequence plays a critical role. The main point of the approach is that before one performs frequency summation the momentum integrals are separable and independent, so the order of momentum integrations is irrelevant [10].

We start from Eq. (5.17):

$$V(R) = \frac{J^2T}{2(2\pi)^s n^2}\sum_m \int d\mathbf{k}d\mathbf{k}' \frac{\exp[i(\mathbf{k}-\mathbf{k}')R]}{(i\omega_m + \varepsilon_{\mathrm{F}} - \varepsilon(\mathbf{k}))(i\omega_m + \varepsilon_{\mathrm{F}} - \varepsilon(\mathbf{k}'))}, \tag{5.27}$$

where $s = 2$ (1D) or 4 (2D), and n is the linear (1D) or sheet (2D) density of the host atoms.

5.4.1 1D- Metal

Interaction Eq. (5.27) contains the product of momentum integrals

$$V^{1D}(R) = \frac{J^2 m^2 T}{2\pi^2 \hbar^4 n^2} \sum_m G(R, i\omega_m) G(-R, i\omega_m),$$

$$G(R, i\omega_m) = -\int_{-\infty}^{\infty} \frac{\exp(ikR)dk}{\left(k^2 - \frac{2im\omega_m}{\hbar^2} - \frac{2m\varepsilon_F}{\hbar^2}\right)}$$

$$= -\int_{\Gamma} \frac{\exp(ikR)dk}{\left(k - \frac{\sqrt{2m}}{\hbar}\sqrt{i\omega_m + \varepsilon_F}\right)\left(k + \frac{\sqrt{2m}}{\hbar}\sqrt{i\omega_m + \varepsilon_F}\right)}, \tag{5.28}$$

where the contour Γ in the complex k-plane is shown in Fig. 5.3.

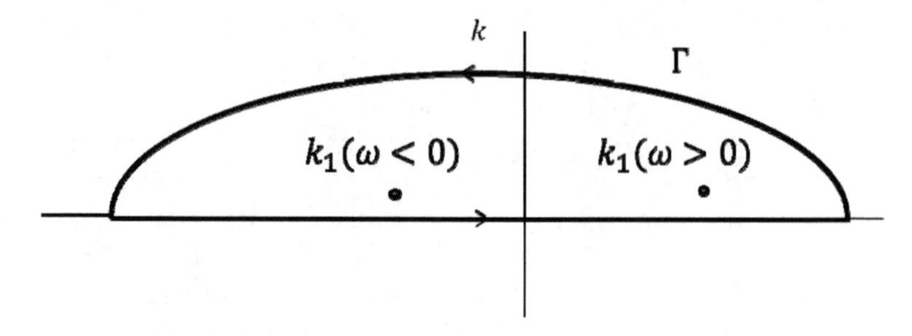

Figure 5.3 Contour and poles in Eq. (5.28).

The contour is closed in the upper k-half-plane where the pole $k_1 = \frac{\sqrt{2m}}{\hbar}\sqrt{i\omega_m + \varepsilon_F}$ is located for any sign of frequency. Evaluating $G(-R, i\omega_m)$ we close the contour in the lower k-half-plane where the pole $k_2 = \frac{\sqrt{2m}}{\hbar}\sqrt{i\omega_m + \varepsilon_F}$ is located. Finally, Eq. (5.28) can be expressed as

$$V^{1D}(R) = -\frac{J^2 mT}{4\hbar^2 n^2} \sum_m \frac{\exp\left[2iRk_F\sqrt{1 + \frac{i\omega_m}{\varepsilon_F}}\right]}{i\omega_m + \varepsilon_F}. \tag{5.29}$$

Now the frequency summation can be replaced by the integral (5.19) for which the integration path is shown in Fig. 5.4.

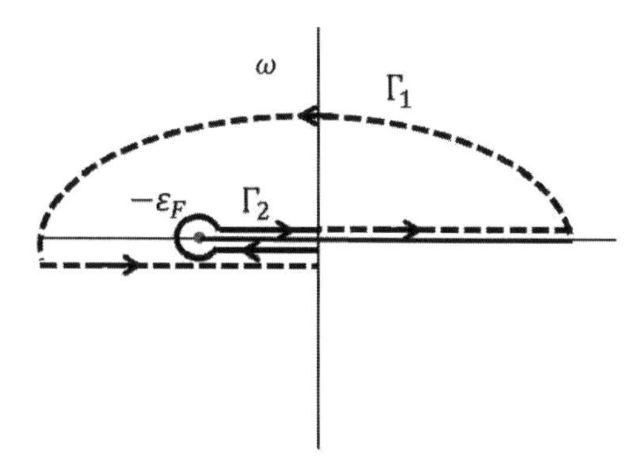

Figure 5.4 Integration path in Eq. (5.29). Contours Γ_1 and Γ_2 are shown as dashed and solid lines, respectively.

Contour Γ_2 goes around the cut $(-\varepsilon_F, \infty)$ and includes a simple pole at $\omega = \varepsilon_F$. Replacing the sum with the integral at $T = 0$ and using the analyticity of the integrand inside the closed path $\Gamma_1 + \Gamma_2$ we come to the integral

$$2\pi i \sum_m g(i\omega_m) = \int_{\Gamma_1} g(\omega)d\omega = -\int_{\Gamma_2} g(\omega)d\omega$$

$$= -\int_0^{-\varepsilon_F} g(\omega)d\omega - \int_{-\varepsilon_F}^{0} g(\omega)d\omega + 2\pi i \, Res[g(\omega)] \qquad (5.30)$$

Integration in first and second terms in (5.30) goes on opposite sides of the cut where the radical in the integrand has opposite signs. The residue in Eq. (5.30) is to be taken at $\omega = -\varepsilon_F$:

$$\sum_m \frac{\exp\left[2iRk_F\sqrt{1+\dfrac{i\omega_m}{\varepsilon_F}}\right]}{i\omega_m + \varepsilon_F}$$

$$= \frac{1}{2\pi i}\int_{-\varepsilon_F}^{0} \frac{\exp\left[-2iRk_F\sqrt{1+\dfrac{\omega}{\varepsilon_F}}\right] - \exp\left[2iRk_F\sqrt{1+\dfrac{\omega}{\varepsilon_F}}\right]}{\omega + \varepsilon_F} d\omega + 1$$

$$= -\frac{1}{\pi}\int_0^1 \frac{\sin[2iRk_F\sqrt{1-y}]}{-y+1} dy + 1 \qquad (5.31)$$

Using the sum (5.31) in Eq. (5.29), we obtain the final indirect exchange interaction:

$$V^{1D}(R) = \frac{J^2 mT}{2\hbar^2 \pi n^2}\left[Si(2k_F R) - \frac{\pi}{2}\right], \tag{5.32}$$

where $Si(x)$ is the sine integral function. Here the second term in Eq. (5.32) which was the subject of the discussion in Refs. [7, 11] appears as a contribution from the pole at $\omega = -\varepsilon_F$.

5.4.2 2D Metal

In a two-dimensional electron gas, the range function (5.27) after integrations over momenta has the form

$$\begin{aligned} V(R) &= \frac{J^2 T}{2n^2}\sum_m I_m(R) I_m(-R), \\ I_m(R) &= \int d\mathbf{k}\,\frac{\exp(i\mathbf{kR})}{i\omega_m + \varepsilon_F - \varepsilon(\mathbf{k})} = \frac{1}{2\pi}\int_0^\infty dk\,\frac{kJ_0(kR)}{i\omega_m + \varepsilon_F - \varepsilon(k)}, \\ J_0(kR) &= \frac{1}{2}[H_0^{(1)}(kR) + H_0^{(2)}(kR)], \end{aligned} \tag{5.33}$$

where $J_0(z)$, $H_\nu^{(n)}(z)$ are Bessel and Hankel functions, respectively. Making use of the analytical continuation [12], $H_0^{(1)}(z) = -H_0^{(2)}(ze^{-i\pi})$, we expand the integration to negative wave vectors shifting the path upward of the real axis thus avoiding the cut $-\infty < z < 0$ (see Fig. 5.5):

$$\begin{aligned} I_m(R) &= \frac{1}{4\pi}\int_0^\infty \frac{[H_0^{(1)}(kR) - H_0^{(1)}(kRe^{i\pi})]kdk}{i\omega_m + \varepsilon_F - \varepsilon(k)} \\ &= \frac{1}{4\pi}\int_0^\infty \frac{H_0^{(1)}(kR)kdk}{i\omega_m + \varepsilon_F - \varepsilon(k)} - \frac{1}{4\pi}\int_0^{-\infty + i\delta} \frac{H_0^{(1)}(kR)kdk}{i\omega_m + \varepsilon_F - \varepsilon(k)} \\ &= \frac{1}{4\pi}\int_\Gamma \frac{H_0^{(1)}(kR)kdk}{i\varepsilon_m + \varepsilon_F - \varepsilon(k)}, \end{aligned} \tag{5.34}$$

where the contour Γ is shown in Fig. 5.5.

The Hankel function is analytical inside the contour and tends to zero on the infinite semicircle, so the integral is equal to $2\pi i * Res$, where *Res* is the residue of the integrand in the pole k_1:

$$I_m(R) = -\frac{im}{2\hbar^2} H_0^{(1)}(Rk_F\sqrt{1 + i\omega_m/\varepsilon_F}), \tag{5.35}$$

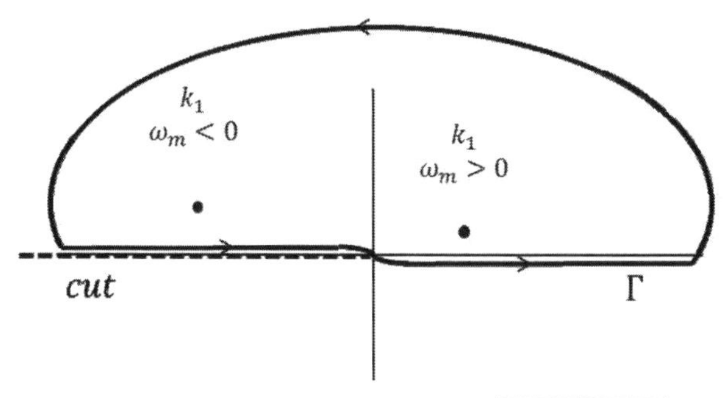

Figure 5.5 Integration path in Eq. (5.34), $k_1 = \sqrt{2m\,(i\omega_m + \varepsilon_F)}$ is the pole of the integrand.

Substituting $I_m(R)I_m(-R)$ into Eq. (5.33) one gets the exchange interaction as a frequency sum

$$V(R) = -\frac{J^2m^2T}{8n^2\hbar^4}\sum_m [H_0^{(1)}(Rk_F\sqrt{1 + i\omega_m/\varepsilon_F})]^2. \tag{5.36}$$

The integration over frequencies by the rule (5.19) goes along the contour shown in Fig. 5.6, which differs from that in Fig. 5.4, as $\omega = -\varepsilon_F$ is just a branch point here, not a pole.

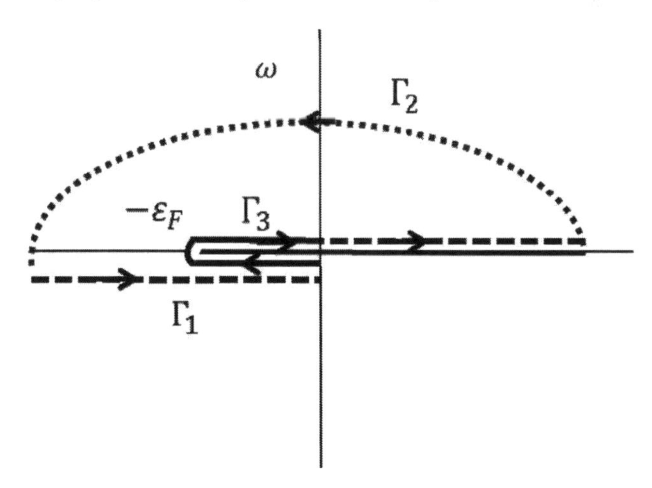

Figure 5.6 Closed integration path for 2D RKKY.

The Hankel function in Eq. (5.36) is analytical inside the contour, $\Gamma_1 + \Gamma_2 + \Gamma_3$, so the integral along the closed path is zero, and finally the two integrals along the path Γ_3 stay with the opposite signs of the square root on both sides of the branch cut:

$$V(R) = -\frac{iJ^2m^2}{16\pi n^2\hbar^4}\left\{\int_{-i\delta}^{-\varepsilon_F - i\delta}\sum_m\left[H_0^{(1)}\left(-Rk_F\sqrt{\frac{1+i\omega_m}{\varepsilon_F}}\right)\right]^2 + \int_{-\varepsilon_F + i\delta}^{i\delta}\sum_m\left[H_0^{(1)}\left(Rk_F\sqrt{\frac{1+i\omega_m}{\varepsilon_F}}\right)\right]^2\right\}$$
$$= -\frac{J^2m}{4\pi n^2\hbar^2R^2}\left\{\int_0^{Rk_F} xJ_0(x)N_0(x)dx\right\}, \tag{5.37}$$

where $J_n(z)$ and $N_n(z)$ are the Bessel and Newmann functions, respectively. In Eq. (5.37) we used the relation between Bessel functions: $H_0^{(1)}(z) = J_0(z) + iN_0(z)$; $H_0^{(2)}(z) = J_0(z) - iN_0(z)$. The integral (5.37) can be evaluated exactly resulting in the final expression for 2D RKKY as follows:

$$V^{2D}(R) = -\frac{J^2mk_F^2}{8\pi^2n^2\hbar^2}\{J_0(Rk_F)N_0(Rk_F) + J_1(Rk_F)N_1(Rk_F)\}. \tag{5.38}$$

The 2D-range function Eq. (5.38) was obtained independently in Refs. [13, 14, 15], and was reproduced here to illustrate the method.

5.5 Exchange Interaction in Semiconductors

In a semiconductor, the chemical potential lies in the bandgap, $\mu = 0$, and there are no carriers in the ground state, $T = 0$. The poles in the integrand of I_1 or I_2 in Eq. (5.20) have the same signs, and then $I_1 = I_2 = 0$. The relevant term in Eq. (5.17) is proportional to $|J|^2I_{12}$. That means the exchange interaction is caused by the interband *s–d* coupling that generates virtual electron–hole pairs excited across the bandgap, as shown in Fig. 5.7. The pairs excited at one localized spin propagate through the crystal until they annihilate at another localized spin.

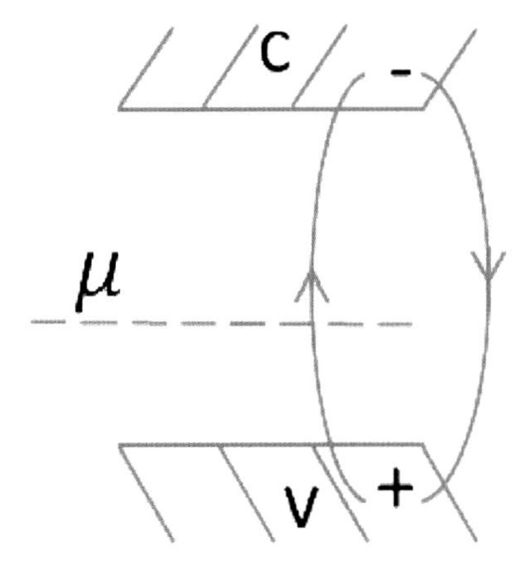

Figure 5.7 Virtual process that creates the Bloembergen–Rowland indirect exchange.

That is the case for Bloembergen–Rowland interaction [16]. Using the energy spectrum from Eq. (5.13) one gets

$$V_{BR}(R) = \frac{J_{12}^2}{N^2}\sum_{\mathbf{k},\mathbf{q}}\frac{1}{2\pi i}\int_\Gamma d\omega\frac{\exp(i\mathbf{qR})}{(\omega-\varepsilon_1(\mathbf{k}))(\omega-\varepsilon_2(\mathbf{k}+\mathbf{q}))}$$

$$= -\frac{mJ_{12}^2}{32n^2\hbar^2\pi^6}\int d\mathbf{k}d\mathbf{k}'\frac{\cos[\mathbf{R}(\mathbf{k}-\mathbf{k}')]}{\mathbf{k}^2+\mathbf{k}'^2+\lambda^2},\ \lambda=\sqrt{\frac{2mE_g}{\hbar^2}}. \quad (5.39)$$

Unlike the one-band RKKY, the momentum integration in (5.40) does not have restrictions:

$$V_{BR}(R) = -\frac{mJ_{12}^2}{4\pi^4n^2\hbar^2R^2}\int_0^\infty kdk\sin(Rk)\int_{-\infty}^{\infty}\frac{k'\sin(Rk')}{k^2+k'^2+\lambda^2}dk'$$

$$= \frac{imJ_{12}^2}{4\pi^4n^2\hbar^2R^2}\int_0^\infty kdk\sin(Rk)\int_{\Gamma_1}\left[\frac{k'\exp(iRk')}{k^2+k'^2+\lambda^2}\right]dk'. \quad (5.40)$$

Contour Γ_1 is shown in Fig. 5.8, $p_{1,2} = \pm i\sqrt{k^2+\lambda^2}$. The negative sign in (5.40) comes from the difference of distribution functions in Eq. (5.22) implying that at $T \to 0$ the conduction band is empty, $f(E_1-\mu)\to 0$, and the valence band is fully occupied, $f(E_2-\mu)\to 1$.

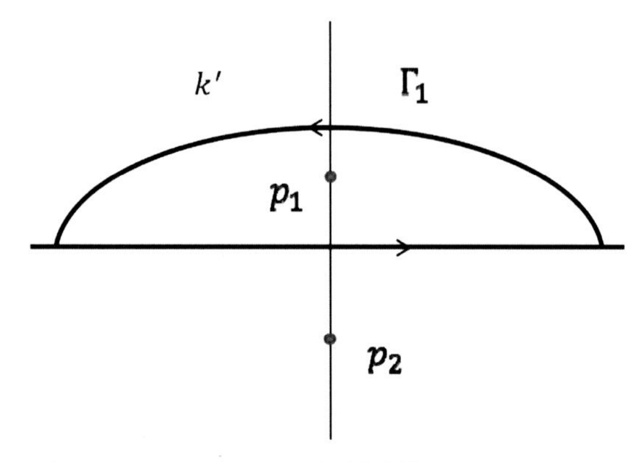

Figure 5.8 Integration contour in Eq. (5.40).

Integration over k' gives $2\pi i$ multiplied by the residue in the pole p_1:

$$V_{BR}(R) = -\frac{m J_{12}^2}{8\pi^3 n^2 \hbar^2 R^4} I,$$
$$I = \int_0^\infty x \sin(x) \exp\left(-\sqrt{x^2 + x_0^2}\right) dx$$
$$x_0 = R/R_0, R_0 = \hbar(2mE_g)^{-1/2}. \qquad (5.41)$$

In order to evaluate the integral we take the derivative $\partial I/\partial x_0$ and find

$$\frac{\partial I}{\partial x_0} = -\frac{x_0^2}{\sqrt{2}} K_1\left(x_0\sqrt{2}\right), \qquad (5.42)$$

where $K_\nu(z)$ is the modified Bessel function. Making use of the relation $\partial[z^\nu K_\nu(z)]/\partial z = -z^\nu K_{\nu-1}(z)$, one gets $I = \frac{1}{2} x_0^2 K_2\left(\sqrt{2}x_0\right)$ and, finally, the interaction takes the form [17]:

$$V_{BR}(R) = -\frac{m J_{12}^2}{16\pi^3 n^2 \hbar^2 R^4} \begin{cases} 1, & R \ll R_0 \\ \frac{\sqrt{\pi}}{2^{3/4}} \left(\frac{R}{R_0}\right)^{3/2} \exp\left(-\frac{\sqrt{2}R}{R_0}\right), & R \gg R_0 \end{cases}. \qquad (5.43)$$

In the long-range limit the interaction (5.43) reproduces the one calculated in Ref. [18], while in the short-range limit it scales as

$1/R^4$ [4]. The exponential decay of the range function is determined by the effective mass and the bandgap. We have considered a simplistic model that keeps effective masses of electrons and holes equal. In a more realistic approach the effective mass in (5.43) should be replaced with the reduced mass: $m \to m_e m_h/(m_e + m_h)$.

Since there are no real carriers to mediate the indirect exchange in a semiconductor at $T = 0$ the Bloembergen–Rowland interaction may be referred to as a lattice contribution to the magnetic interaction in dielectrics. Interaction $V_{BR}(R)$ does not contain oscillations obtained in the original paper [16]. These oscillations appear as a result of cutting the integral over momentum on the upper limit, and as it was pointed out in Ref. [18], they cannot be considered to be real.

5.6 Indirect Magnetic Exchange through the Impurity Band

Recent experimental data show that ferromagnetism in GaAs(Mn) exist in highly resistive samples where the chemical potential lies in the bandgap close to or inside the Mn-related impurity band (IB), so the Fermi surface does not exist [1]. Also, one could not apply the RKKY-interaction to a wide bandgap AlGaN-based material system where degenerate carriers could hardly exist either in conduction or valence bands. Other sources of exchange interaction associated with IB might explain ferromagnetic ordering in low carrier density GaAs structures as well as in wide bandgap GaN-related DMS. One of them is the double-exchange. While the RKKY model implies that s–d or p–d interactions between the spin of the carrier and the localized spin is much less than the electron kinetic energy, i.e., s–d exchange is small as compared to the width of conduction (s–d) or valence (p–d) bands, the double-exchange model explores the opposite limit: s–d interaction is much stronger than the width of the band. That's why the model can be applied to the narrow IB. The Hamiltonian of the model reads

$$H = -\sum_{ijs} t_{ij} C_{is}^{+} C_{js} - J\sum_{i} \mathbf{s}_i \mathbf{S}_i, \quad (5.44)$$

where the first term is the kinetic energy of the electrons jumping from site i to site j, the second term is the s–d exchange, $J >> zt$, z is the number of nearest neighbors. The model, solved by Anderson and Hasegawa [20] and deGennes [21], shows that the energy of low-energy excitations can be described by an effective jumps: $\tilde{t}_{ij} = t\cos(\alpha_{ij}/2)$, α is the angle between neighbor spins $\mathbf{S}_i$ and $\mathbf{S}_j$. This means that the minimum energy corresponds to electron jumps between sites with parallel spins, in other words, delocalized electrons in IB favor the localized spins to align in parallel. The model works if the chemical potential lies inside the IB; otherwise IB is fully occupied or completely empty and therefore does not contribute to the exchange interaction.

There is another IB-related mechanism of indirect exchange that works even if the IB is fully occupied or completely empty [22]. The mechanism is similar to the Bloembergen–Rowland one (see Section 5.5) where one of participating bands is the IB.

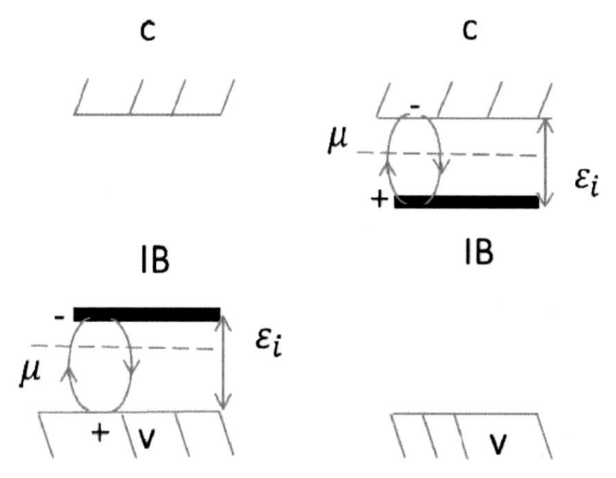

Figure 5.9 Virtual processes responsible for the impurity band related indirect exchange [22].

The basic approach is illustrated in Fig. 5.9. Since the effective mass in a narrow IB is much larger than in the c, v-bands, the analytical interaction can be obtained from Eq. (5.43), where the reduced effective mass can be approximately replaced with an effective mass of the corresponding c- or v-band. The essential difference between the IB-related mechanism and Bloembergen–Rowland interaction is in the radius of interaction. In III-V semiconductors, the Bloembergen–Rowland contribution is negligible as the radius R_0 (5.41) is determined by a large bandgap, making the interaction negligibly small at a distance larger than the lattice constant. The IB-related mechanism relies on a smaller gap ε_i instead (see Fig. 5.9):

$$V_{IB}(R) = -\frac{m_v J_{12}^2}{16\pi^3 n^2 \hbar^2 R^4} \begin{cases} 1, & R \ll R_{ib} \\ \dfrac{\sqrt{\pi}}{2^{3/4}}\left(\dfrac{R}{R_{ib}}\right)^{3/2} \exp(-\sqrt{2}\,R/R_{ib}), & R \gg R_{ib} \end{cases}$$

$$R_{ib} = \hbar(2m_v\varepsilon_i)^{-1/2}. \tag{5.45}$$

Since $R_{ib} >> R_0$, the interaction through the impurity band $V_{IB}(R)$ is much more long-ranged than the Bloembergen–Rowland interaction and could be quantitatively comparable to RKKY in metals.

This interaction has been used to explain ferromagnetic phase transition in GaAsMn in the regime when the mean-field theory is not applicable (see Chapter 6) [22].

5.7 Conclusions

A localized spin excites band electrons due to $s(p) - d$ exchange interaction and generally gives rise to three types of indirect exchange interaction caused by virtual electron–hole excitations. If the chemical potential lies in the energy gap, there is a threshold for electron–hole excitations that translates to an exponential drop of the range function. The energy gap determines the characteristic length of the exponential decay (Bloembergen–Rowland and impurity band related mechanisms). In indirect-gap semiconductors the electron–hole excitations involve the momentum transfer, $\mathbf{K}$. This modulates the exponential decay by oscillations with a period $\sim|\mathbf{K}|^{-1}$ [18]. If the chemical potential lies inside the band and the Fermi level exists the gapless electron–hole excitations near k_F result in the range function that oscillates with a period $\sim k_\mathrm{F}^{-1}$ and an amplitude falling as the power of the distance. That is long-range RKKY interaction.

It is instructive to look at RKKY range functions at large distances, $k_\mathrm{F}R >> 1$, especially in one and two dimensions. In a 1D metal, the large-distance decay of the range function is provided by the second term in Eq. (5.32). In a 2D metal, the asymptotic terms proportional to R^{-1} cancel each other in Eq.(5.38) and the leading term is proportional to R^{-2}. For all dimensions, the large-distance asymptotic expansions in Eqs. (5.24), (5.32), and (5.38) give

$$V^{1D}(R) \sim \frac{-\cos(2Rk_F)}{R};\ V^{2D}(R) \sim \frac{-\sin(2Rk_F)}{R^2},\ V_{RKKY}(R) \sim \frac{-\cos(2k_F R)}{R^3} \tag{5.46}$$

The amplitude of range oscillations decreases following the power law $R^{-\alpha}$, where α is the spatial dimension.

Problems

5.1 Find the long-range asymptotic of RKKY interaction in two dimensions.

5.2 Indirect exchange interaction is mediated by conduction electrons with parabolic spectrum. Electron density, electron effective mass, and mean free path are $n = 10^{19}$ cm^{-3}, $m = 0.067\ m_0$, and $l = 10$ Å, respectively. The sample is doped with magnetic impurities to the extent that the average distance between them is $R_{av} = 20$ Å. Do magnetic atoms feel oscillatory indirect exchange interaction?

References

1. Tanaka M, Ohya S, Hai PN (2014) Recent progress in III-V based ferromagnetic semiconductors: Band structure, Fermi level, and tunneling transport, *Appl Phys Rev*, **1**, 011102.
2. Liu C, Yun F, Morkoç H (2005) Ferromagnetism of ZnO and GaN: A review, *J Mater Sci Mater Electron*, **16**, 555–597.
3. Abrikosov AA, Gorkov LP, Dzyaloshinski IE (1965) *Quantum Field Theoretical Methods in Statistical Mechanics*, Pergamon, New York.
4. Ruderman MA, Kittel C (1954) Indirect exchange coupling of nuclear magnetic moments by conduction electrons, *Phys Rev*, **96**, 99–102; Kasuya T (1956) A Theory of metallic ferro- and antiferromagnetism on Zener's model, *Prog Theor Phys*, **16**, 45–57; Yosida K (1957) Magnetic properties of Cu-Mn alloys, *Phys Rev*, **106**, 893–898.
5. De Gennes PG (1962) Polarisation de charge (ou de spin) au voisinage d'une impureté dans un alliage, *J Phys Radium*, **23**, 630–636.
6. Mattis DC (2006) *The Theory of Magnetism Made Simple*, World Scientific, Singapore.

7. Yafet Y (1987) Ruderman-Kittel-Kasuya-Yosida range function of a one-dimensional free-electron gas, *Phys Rev B*, **36**, 3948–3949.
8. Dugaev VK, Litvinov VI (1990) Low-temperature spin glass in IV-VI semimagnetic semiconductors, *Phys Rev B*, **41**, 788–790; Dugaev VK, Litvinov, VI, Petrov PP (1994) Magnetic impurity interactions in a quantum well on the base of IV-VI semiconductors, *Superlattices Microstr*, **16**, 413–417.
9. Aristov DN (1996) Indirect RKKY interaction in any dimensionality, *Phys Rev B*, **55**(13), 8064–8066.
10. Litvinov VI, Dugaev VK (1996) RKKY interaction in one-and two-dimensional electron gases, *Phys Rev B*, **58**(7), 3584–3585.
11. Kittel C (1968), in *Solid State Physics* (Zeitz F, Turnbull D, Ehreinreich H, ed), Academic, New York, vol. 22, pp. 360–366, see also erratum.
12. Abramowitz M, Stegun IA (eds) (1964) *Handbook of Mathematical Functions*, National Bureau of Standards, New York.
13. Korenblit IY, Shender EF (1975) Dilute ferromagnetic alloys with long-range exchange interaction, *Sov Phys JETP*, **42**, 566–569.
14. Fischer B, Klein MW (1975) Magnetic and nonmagnetic impurities in two-dimensional metals, *Phys Rev B*, **11**, 2025–2029.
15. Beal-Monod MT (1987) Processes yielding high superconducting temperatures, *Phys Rev B*, **36**(16), 8788–8790.
16. Bloembergen N, Rowland TJ (1955) Nuclear spin exchange in solids: Tl_{203} and Tl_{205} magnetic resonance in thallium and thallic oxide, *Phys Rev*, **97**(6), 1680–1698.
17. Litvinov VI (1985) Indirect exchange interaction between magnetic centers in a narrow-gap semiconductor, *Sov Phys Semicond*, **19**, 345–346.
18. Abrikosov AA (1980) Spin-glass with a semiconductor as host, *J Low Temp Phys*, **39**(1/2), 217–229.
19. Liu L, Bastard G (1982) Indirect exchange interaction in lead salts, *Phys Rev B*, **25**(1), 487–489.
20. Anderson PW, Hasegawa H (1955) Considerations on double exchange, *Phys Rev*, **100**, 675–681.
21. De Gennes PG (1960) Effects of double exchange in magnetic crystals, *Phys Rev*, **118**, 141–154.
22. Litvinov VI, Dugaev VK (2001) Ferromagnetism in magnetically doped III-V semiconductors, *Phys Rev Lett*, **86**(24), 5593–5596.

Chapter 6

Ferromagnetism in III-V Semiconductors

Diluted magnetic semiconductors, namely, semiconductors doped with magnetic impurities, or ternary alloys with transition metal (TM) components (Mn, Fe, Co) are recognized as templates for various devices in semiconductor spintronics. In this chapter we briefly discuss the physics related to the magnetic properties of wideband and narrow-gap III-V DMS bearing in mind the main subject of theoretical and experimental activity—room-temperature ferromagnetism. Material science aspects of ferromagnetism include the growth method, the choice of magnetic component and its content, and intentional manipulation of the electron spectrum of the host matrix by fabricating artificial structures like quantum wells and quantum dots.

Experimental Curie temperatures and various theoretical approaches to ferromagnetic phase transitions in narrow-gap GaAs(TM) and wideband GaN(TM) have been topics of discussion for a long time now. The explanation of ferromagnetic phase transition in GaAs(TM) and GaN (TM) is normally based on the mean-field approximation (MFA) applied to the RKKY interaction (or its modification) [1–3]. It is instructive to look at the theory of ferromagnetism in DMS when MFA validity is under the question and to find an adequate picture of phase transition that can explain experimental results obtained to date. In this chapter, we

Wide Bandgap Semiconductor Spintronics
Vladimir Litvinov

ISBN 978-981-4669-70-2 (Hardcover), 978-981-4669-71-9 (eBook)
www.panstanford.com

discuss some features of the theory when the MFA is hardly or not at all applicable to GaAs- and GaN-based DMS.

6.1 Mean-Field Approximation

Interacting spins in a magnetic field are described by the isotropic Heisenberg Hamiltonian

$$H = g\mu_B \sum_i \mathbf{B}\hat{\mathbf{S}}_i - \frac{1}{2}\sum_{i,j} V(\mathbf{r}_i - \mathbf{r}_j)\hat{\mathbf{S}}_i\hat{\mathbf{S}}_j, \tag{6.1}$$

where $\mathbf{B}$ is the external field magnetic induction, $V(\mathbf{R})$ is the exchange interaction, g and μ_B are the magnetic impurity g-factor and the Bohr magneton, respectively. The first term in Eq. (6.1), the Zeeman energy, is written in a simplified form that does not take into account the orbital momentum of a magnetic atom. We will come back to this point in more detail in Section 6.3.1 where the g-factor is discussed.

It is convenient to represent the spin operator in (6.1) as the sum of thermodynamic average $\eta = \langle\hat{\mathbf{S}}_i\rangle$ and fluctuation $(\hat{\mathbf{S}}_i - \eta)$. Mean-field approximation, proposed by Weiss, neglects fluctuations, so the interaction term in (6.1) can be expressed as

$$-\frac{1}{2}\sum_{i,j} V(\mathbf{r}_i - \mathbf{r}_j)\hat{\mathbf{S}}_i\hat{\mathbf{S}}_j \rightarrow -\sum_i \hat{\mathbf{S}}_i \sum_j V(\mathbf{r}_i - \mathbf{r}_j)\eta_j + \frac{1}{2}\sum_{ij} V(\mathbf{r}_i - \mathbf{r}_j)\eta_i\eta_j. \tag{6.2}$$

The approximation decouples the interaction term in (6.1) replacing it with a term describing spins in an effective molecular field $\mathbf{B}_{mi}$

$$H = g\mu_B \sum_i (\mathbf{B} + \mathbf{B}_{mi})\hat{\mathbf{S}}_i + \frac{1}{2}\sum_{ij} V(\mathbf{r}_i - \mathbf{r}_j)\eta_i\eta_j,$$

$$\mathbf{B}_{mi} = -\frac{1}{g\mu_B}\sum_j V(\mathbf{r}_i - \mathbf{r}_j)\eta_j \tag{6.3}$$

Ferromagnetic ordering means that in the ground state local moments are aligned parallel to each other. As we consider the isotropic exchange interaction, the alignment takes place along the external magnetic field axis Z, i.e., $\eta = (0, 0, \eta)$. The average spin on a site does not depend on the site position, $\mathbf{r}_i$: $\eta_i = \eta$. The MFA Hamiltonian (6.3) can be expressed in terms of one-site magnetic moment $M = -g\mu_B\eta$ as follows:

$$H = g\mu_B(B_z + \eta V_0)\sum_{i=1}^{N_m} S_{iz} + \frac{1}{2}N_m V_0 \eta^2 \equiv b\sum_{i=1}^{N_m} S_{iz} + qN_m.$$

$$V_0 = \sum_j V(\mathbf{r}_i - \mathbf{r}_j);\ b = g\mu_B B_{\text{eff}};\ B_{\text{eff}} = B_z + \frac{V_0 M}{(g\mu_B)^2};\ q = \frac{1}{2}\frac{V_0 M^2}{(g\mu_B)^2}, \qquad (6.4)$$

where N_m is the number of magnetic sites. In a periodic magnetic lattice the interaction parameter V_0 is the zero-Fourier transform of the range function. As the range function descreases with distance, the parameter can be approximated by contributions from nearest magnetic neighbors only: $V_0 = zV$, where z is the number of nearest neighbors or the coordination number of the lattice. The effective magnetic field B_{eff} acting on localized spin is the self-consistent field that includes the external field and the field originating from surrounding magnetic moments. The magnetic moment is found from thermodynamic considerations. Thermodynamic quantities can be obtained from the partition function

$$Z = Tr[\exp(-\beta H)] = Tr\left[\exp\left\{-\beta b\sum_{i=1}^{N_m} S_{iz} - \beta q\sum_{i=1}^{N_m} 1\right\}\right]$$

$$= (2S+1)^{N_m}\ \exp(-\beta q N_m) * Tr\left\{\exp\left[-\beta b\sum_{i=1}^{N_m} S_{iz}\right]\right\}$$

$$= (2S+1)^{N_m}\ \exp(-\beta q N_m)\prod_{i=1}^{N_m}\sum_{n=-S}^{S}\exp(-\beta b n)$$

$$= (2S+1)^{N_m}\ \exp(-\beta q N_m)\left\{\sum_{n=-S}^{S}\exp(-\beta b n)\right\}^{N_m}$$

$$= (2S+1)^{N_m}\ \exp(-\beta q N_m)\left[\frac{\sinh((S+1/2)\beta b}{\sinh(\beta b/2)}\right]^{N_m};\ \beta = \frac{1}{T}. \qquad (6.5)$$

The trace operator in (6.5) performs summation over $2S + 1$ spin states (eigenvalues of S_z) that is calculated as a geometric series. Using Eq. (6.5) one can express the Helmholtz free energy per magnetic site as

$$\frac{F}{N_m} = -T\log Z$$

$$= q - T\log(2S+1) - T\log\frac{\sinh\left(\left(S+\frac{1}{2}\right)\beta b\right)}{\sinh\left(\frac{\beta b}{2}\right)}; \tag{6.6}$$

The magnetic moment m in thermodynamic equilibrium can be found from the minimum free energy, $\partial F/\partial M = 0$:

$$\frac{\partial F}{N_m \partial M} = \frac{V_0 M}{(g\mu_B)^2} - T\frac{\sinh\left(\frac{\beta b}{2}\right)}{\sinh\left(\left(S+\frac{1}{2}\right)\beta b\right)}\frac{\partial}{\partial M}\frac{\sinh\left(\left(S+\frac{1}{2}\right)\beta b\right)}{\sinh\left(\frac{\beta b}{2}\right)}$$

$$= \frac{V_0 M}{(g\mu_B)^2} - \frac{V_0}{g\mu_B}S\left\{\frac{2S+1}{2S}\coth\left[\left(S+\frac{1}{2}\right)\beta b\right] - \frac{1}{2S}\coth\left[\frac{\beta b}{2}\right]\right\};$$

$$\frac{\partial F}{\partial M} = 0 \rightarrow M = g\mu_B S B_S\left(\beta S\left(g\mu_B B_z + \frac{V_0 M}{g\mu_B}\right)\right), \tag{6.7}$$

where $B_S(x)$ is the Brillouin function:

$$B_S(x) \equiv \frac{2S+1}{2S}\coth\left[\left(1+\frac{1}{2S}\right)x\right] - \frac{1}{2S}\coth\left[\frac{x}{2S}\right]. \tag{6.8}$$

The function varies from -1 to 1 when x runs from $-\infty$ to $+\infty$.

If an external magnetic field is zero, the M-dependence of the left and right sides of Eq. (6.7) are shown in Fig. 6.1. Equation (6.7) has three solutions, one of which is trivial, $M_0 = 0$. Non-trivial solutions $\pm M_0$ are indicated in Fig. 6.1 as intersections of the graphs. They exist only if the slope of $SB_S(M)$ at $M \rightarrow 0$ is larger than unity:

$$\frac{\partial}{\partial M}\left\{SB_S\left(\beta S\frac{V_0 M}{g\mu_B}\right)\right\} > 1 \tag{6.9}$$

Making use of the expansion of the Brillouin function at $x \rightarrow 0$: $B_S(x) \rightarrow [(S+1)/3S]x$, the condition (6.9) can be rewritten as

$$T < T_c,\ T_c = \frac{S(S+1)V_0}{3}, \tag{6.10}$$

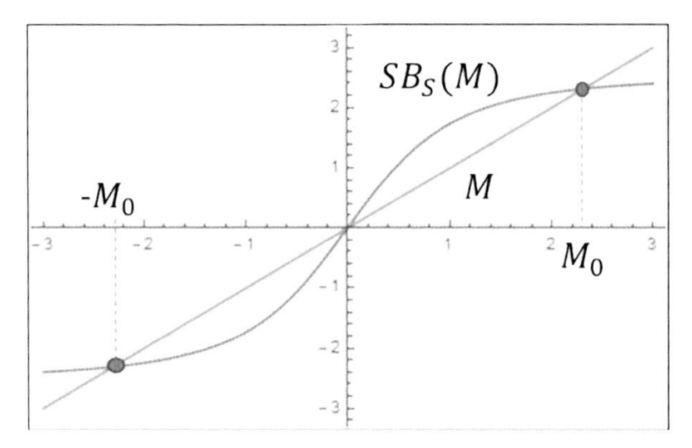

Figure 6.1 Graphic solutions to Eq. (6.7), $S = 5/2$.

where T_c is the critical temperature of the ferromagnetic phase transition.

To confirm that non-trivial solutions deliver stable thermodynamic equilibrium we inspect the second derivative of the free energy (6.6):

$$V_0\left[1 - \frac{g\mu_B S\partial B_S}{\partial M}\right] > 0. \tag{6.11}$$

Inequality (6.11) warrants that the free energy has minima at $M = \pm M_0$. The condition holds if $V_0 > 0$ The positive sign of the square bracket in (6.11) is clearly seen in Fig. 6.1 as the slope of $SB_S(M)$ is less than unity at the intersection points. The magnetic moment in the paramagnetic phase $(T \to T_c^+)$ is proportional to B_z. It can be calculated using Eqs. (6.7), (6.10) and expanding the Brillouin function in the limit $M \to 0$, $B_z \to 0$:

$$M = \frac{(g\mu_B)^2}{3}\frac{S(S+1)}{T - T_c}B_z \tag{6.12}$$

Magnetic susceptibility follows from (6.12) as the Curie–Weiss law:

$$\chi = N_m\frac{\partial M}{\partial B_z} = \frac{N_m(g\mu_B)^2}{3}\frac{S(S+1)}{T - T_c}, T \geq T_c. \tag{6.13}$$

We reduced the rotational symmetry of the initially isotropic Heisenberg model to the discrete symmetry of the Ising model in Eq. (6.4). In the absence of an external magnetic field this results in twofold degeneracy ($\pm M_0$) of the ferromagnetic state. As the direction of the external field can be chosen arbitrarily, the derivation of the self-consistent magnetic moment in the Weiss approximation (6.7) can also be applied to the Heisenberg model. In that case, with no external field, the ferromagnetic state is continuously degenerate with respect to three-dimensional rotations of the magnetic moment.

Let us look at the free energy (6.6) shown in Fig. 6.2a as a function of total magnetic moment M. At $B_z = 0$, below the critical temperature the free energy has two minima that correspond to the equilibrium values of the magnetic moment $\pm M_0$. This two-fold degeneracy is broken by an external magnetic field as seen in Fig. 6.2b. This means that only one direction (linked to the external field) of the magnetic moment corresponds to the absolute minimum of the free energy.

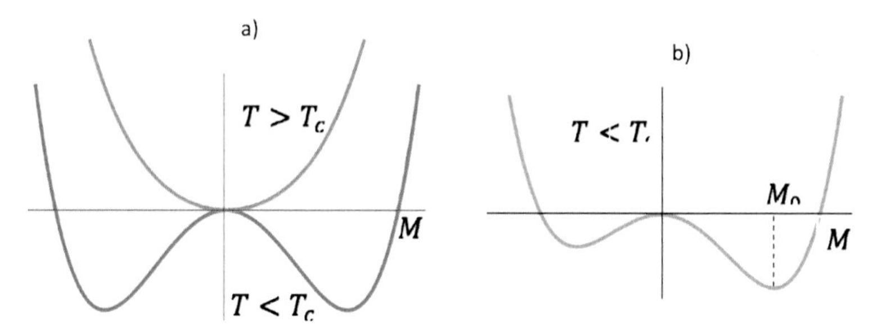

Figure 6.2 (a) Free energy at zero external magnetic field, $B_z = 0$, (b) $B_z \neq 0$.

If we keep a ferromagnetic at $T < T_c$ and remove the external field, $B_z \to 0$ the system preserves magnetic moment M_0 as equilibrium moment because in order to change the moment direction one has to pay the energy cost needed to overcome the potential barrier between the two minima. That is what is called spontaneously broken symmetry.

MFA is valid if the range of exchange interaction is large enough to provide for a strong molecular field created by a large number of magnetic neighbors. Both the interaction range and the coordination number of the lattice contribute to the effective field that suppresses fluctuations and align all spins in parallel. Fluctuations become more noticeable in low-dimensional systems. So, in one-

and two-dimensional magnetic structures the ferromagnetic long-range order does not exist if the interaction is short-ranged (Mermin–Wagner theorem).

When applying the MFT to DMS we deal with magnetic impurities placed randomly throughout the host so it is necessary to check the validity of the approximation because it fails if the interaction radius is equal to or shorter than the average inter-spin distance. There is one more situation where the MFT needs corrections. MFT implies that all localized spins in the lattice have a definite valence state and thus the same degeneracy factor, namely, $J(J + 1)$, J being the value of total angular momentum. It is a good approximation if the Fermi level ε_F lies far enough from the localized level ε_d created by a magnetic center, $|\varepsilon_F - \varepsilon_d| >> T$. However, if it is not a case, the mixed valence of the magnetic impurity has to be taken into account as different valence states contribute differently into the total magnetic moment of the crystal and the self-consistent equation (6.7) should be modified. Below we consider several situations where MFA either fails completely or should be modified to describe the experimental conditions. We also discuss an alternative to the MFT picture of the ferromagnetic phase transition in III-V DMS when MFT is not applicable.

6.2 Percolation Mechanism of the Ferromagnetic Phase Transition

It is often implied that ferromagnetism in GaAs(Mn, Fe) originates from the carrier-mediated exchange interaction (see Chapter 5, Eq. (5.24)) and can be described within the MFA [2]. This approach is valid under two conditions. First, the sample should be degenerate, i.e., the carrier density is high enough to create a Fermi surface in the valence band. Otherwise, one could not use RKKY where the Fermi momentum enters as a parameter. Second, the interaction radius is much larger than the average inter-spin distance in order to create a strong molecular field from the magnetic neighbors. Otherwise, the mean-field theory is not justified.

In GaInAs(Mn) the effective mass of electrons and holes is small enough to help create degenerate holes by Mn- and non-magnetic co-doping. If the average distance between Mn atoms is less than the inverse Fermi wavenumber k_F^{-1} the RKKY interaction favors ferromagnetism and at a range $Rk_F << 1$ follows from that calculated in Chapter 5, Eq. (5.26):

$$\tilde{V}(R) = -\frac{m J_1^2 k_F}{32\, n^2 \pi^3 \hbar^2\, R^3} \exp(-R/l). \tag{6.14}$$

The parameter V_0 that determines the mean-field critical temperature can be calculated by summing up the contributions coming from magnetic neighbors:

$$T_c = \frac{S(S+1)}{3} x \sum_r z_r \tilde{V}(r) \tag{6.15}$$

where z_r is the number of cation sites on a sphere of radius r, x is the Mn content, i.e., the fraction of cation sites filled with Mn atoms. As follows from (6.15), the mean-field carrier-induced mechanism results in $T_c \sim xk_F$. Since $k_F \sim p^{1/3}$ being the hole density in the valence band, the critical temperature follows the trend $T_c \sim xp^{1/3}$ which is close to the one experimentally observed in GaAs(Mn). If Mn content exceeds $x = 0.06$ defects compensate Mn-related p-doping, thus suppressing the hole density and T_c.

Let us discuss situations to which the mean-field model is not applicable. The interaction (6.14) is short-ranged with the radius restricted by crystal imperfections and determined by the carrier mean free path l. In typical experimental samples $l \approx (5 - 6.5)$ Å [4]. The Mn content $x = 0.053$ corresponds to the average inter-spin distance of $\bar{R} = 6$ Å. The main condition for the mean-field approximation is that the local effective magnetic field acting on a magnetic atom is caused by a large number of nearest magnetic atoms. In other words, the interaction radius is supposed to be much larger than the average inter-spin distance, $\bar{R} << l$. This condition does not hold, and the phase transition is better described by the percolation approach [5].

The percolation picture of phase transition deals with magnetic clusters. When the temperature becomes as low as the exchange interaction between a pair of closely located spins, the pair forms a ferromagnetic cluster. The size of the cluster R_{cl} is the solution of the equation

$$T \approx \frac{S(S+1)}{3} \tilde{V}(R_{cl}). \tag{6.16}$$

The cluster size depends on temperature. When temperature decreases, the clusters grow in size due to coalescence (see Fig. 6.3), and the percolation threshold is reached when an "infinite" cluster penetrates the whole sample. This happens at $T = T_c$.

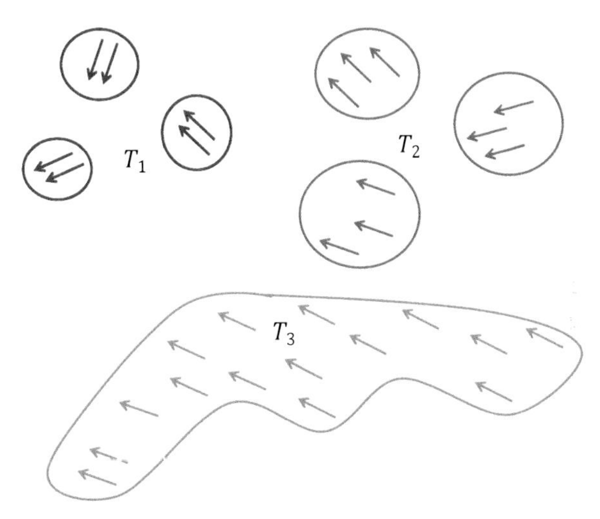

Figure 6.3 Magnetic cluster coalescence, $T_1 > T_2 > T_3 > T_c$.

In the three-dimensional random-sphere model the percolation threshold is reached when the cluster size becomes $R_{\text{perc}} = \bar{R} B_c^{1/3}$:

$$T_c = \frac{S(S+1)}{3} \tilde{V}(R_{\text{perc}}), \tag{6.17}$$

where $B_c = 2.4$ is the geometrical factor. The average inter-spin distance in the cation sublattice of III-V crystals relates to the Mn content. Unit cell volume in GaAs $V_{\text{cell}} = a/4$ contains one cation site, so the Mn content is $x = N_{\text{Mn}}/N_{\text{cell}}$, where N_{Mn}, N_{cell} are the number of magnetic atoms and the number of unit cells, respectively. Average inter-spin distance is expressed as

$$\frac{4}{3}\pi\left(\frac{\bar{R}}{2}\right)^3 N_{\text{Mn}} = V_{\text{cell}} N_{\text{cell}} \rightarrow \bar{R}_{\text{GaAs}} = a\left(\frac{3}{8\pi x}\right)^{1/3}. \tag{6.18}$$

In wurtzite GaN-based crystals the calculation of the average distance differs from Eq. (6.18) as the unit cell of volume $V_{\text{cell}} = a^2 c\sqrt{3}/2$ contains two cation atoms: $2x = N_{\text{Mn}}/N_{\text{cell}}$:

$$\overline{R}_{\text{GaN}} = \left(\frac{3a^2c}{8\pi x}\right)^{1/3}, \tag{6.19}$$

where a, c are the lattice constants in the basal plane and along the hexagonal axis, respectively.

Finally, Eqs. (6.17), (6.18), and (6.19) determine the ferromagnetic critical temperature. Unlike the mean-field theory, the percolation mechanism results in T_c which is non-linear function of x. Examples of how percolation is used in the explanation of the phase transition in GaAs and GaN can be found in Ref. [6].

6.3 Mixed Valence and Ferromagnetic Phase Transition

The doping of GaAs and GaN with transition metals [1,7–9] creates impurity bands in the bandgap. If the co-doping with electrically active non-magnetic impurities and growth-related vacancies in cation and (or) anion sub-lattice shift the chemical potential close to the magnetic impurity band, this makes the valence state of the magnetic impurities variable with temperature and doping. At $T = 0$ the total angular moment of a magnetic impurity changes when the chemical potential crosses the impurity level. At finite temperatures the thermodynamic equilibrium determines the number of magnetic centers with different total momenta and g-factors if the chemical potential lies close to the magnetic impurity level, $|\mu - \varepsilon_i| \leq T$. Variable magnetic moments affect the magnetic susceptibility and the ferromagnetic critical temperature. Below we describe this type of phase transition in more detail.

6.3.1 Magnetic Moment

Let us take a look at how the magnetic moment of the magnetic center relates to its total angular momentum. The Zeeman term in Eq. (6.1) that describes the interaction of a magnetic atom with an external magnetic field contains the operator of localized spin. It is correct until we deal with the local spin created by atomic s-electrons which do not carry an orbital momentum, $L = 0$. Spin moments coming from transition metals and rare earth elements originate from d- and f- atomic shells carrying finite orbital moments that contribute to the total magnetic moment of a magnetic impurity. Formally, this contribution can be obtained from the Hamiltonian of an atomic electron in a constant magnetic field

$$H = -\mu_B \hat{\mathbf{S}}\mathbf{B} + \frac{(\hat{\mathbf{p}} - e\hat{\mathbf{A}})^2}{2m} + eU(\mathbf{r}), \tag{6.20}$$

where $\mu_B = e\hbar/2m$; $\hat{\mathbf{A}} = \mathbf{B} \times \hat{\mathbf{r}}/2$ is the vector-potential, $U(r)$ is the atomic electrostatic potential. The first term in Eq. (6.20) is equivalent to that in Eq. (6.1). The essential difference between (6.1) and (6.20) is that the Hamiltonian (6.20) contains an additional contribution that originates from the kinetic energy (second term). Substituting $\hat{\mathbf{A}}$ in (6.20), one gets the part of the Hamiltonian proportional to the magnetic field

$$H_{Zeeman} = -\mu_B(\hat{\mathbf{J}} + \hat{\mathbf{S}})\mathbf{B};$$
$$\hat{\mathbf{J}} = \hat{\mathbf{L}} + \hat{\mathbf{S}};\ \hat{\mathbf{L}} = \frac{1}{\hbar}[\hat{\mathbf{r}} \times \hat{\mathbf{p}}], \tag{6.21}$$

where $\hat{\mathbf{L}}$ and $\hat{\mathbf{J}}$ are operators of orbital momentum and total angular momentum, respectively.

If the external magnetic field is absent, the state of the atom is degenerate over the directions of the total momentum. The magnetic field lifts the degeneracy and sets the multiplet corresponding to $2J + 1$ eigenvalues of $\hat{J}_z$, z being the axis parallel to the magnetic field. Since the total momentum is the only conserved quantity, directions other than $\hat{\mathbf{J}}$ are undefined, so the average value of $\hat{\mathbf{S}}$ should be parallel to $\hat{\mathbf{J}}$ as all other components are averaged out due to precession around the vector $\hat{\mathbf{J}}$. It is convenient to represent the Zeeman term (6.21) as a term proportional to the total angular momentum:

$$\hat{\mathbf{J}} + \hat{\mathbf{S}} = \hat{G}\hat{\mathbf{J}} \tag{6.22}$$

Scalar multiplication of the left and right sides of Eq. (6.22) by vector $\hat{\mathbf{J}}$ yields

$$\hat{G} = 1 + \frac{\hat{\mathbf{J}}\hat{\mathbf{S}}}{\hat{\mathbf{J}}^2} = 1 + \frac{\hat{\mathbf{J}}^2 + \hat{\mathbf{S}}^2 - \hat{\mathbf{L}}^2}{2\hat{\mathbf{J}}^2}. \tag{6.23}$$

Within the first-order perturbation approach, the magnetic field-induced splitting of the energy level is determined by the matrix element of H_{Zeeman} on atomic wave functions (**B** is directed along the z-axis):

$$\Delta E = \mu_B \left\langle jlsm_J \,|\hat{G}\hat{J}_z|\, jlsm_J \right\rangle B_z, \tag{6.24}$$

where j, l, s, m_J are the quantum numbers that characterize the electron state in a spherically symmetric atomic potential and run over eigenvalues of $\hat{\mathbf{J}}^2$, $\hat{\mathbf{L}}^2$, $\hat{\mathbf{S}}^2$, and $\hat{J}_z$ respectively. States $\left|jlsm_J\right\rangle$ are the egenfunctions of operators $\hat{G}$ and $\hat{J}_z$, so in the matrix elements the operators can be replaced by their eigenvalues:

$$\left\langle jlsm_J \,|\hat{G}|\, jlsm_J \right\rangle = 1 + \frac{J(J+1) + S(S+1) - L(L+1)}{2J(J+1)}$$

$$\left\langle njlm_J \,|J_z|\, njlm_J \right\rangle = m_J,\ m_J = -J, -J+1, \ldots J, \tag{6.25}$$

where J, L, S are the values of total, orbital, and spin angular momentum, respectively. Finally, the Zeeman effect, the magnetic field splitting of energy levels (6.24) is written as

$$\Delta E = g\mu_B B_z m_J,\ g = 1 + \frac{J(J+1) + S(S+1) - L(L+1)}{2J(J+1)}, \tag{6.26}$$

and the absolute value of the magnetic moment is determined by the value of total angular momentum

$$M = g\mu_B|\mathbf{J}| = g\mu_B\sqrt{J(J+1)}. \tag{6.27}$$

If the localized spin is formed by valence s-electrons, then $L = 0$, $g = 2$. The total Hamiltonian (6.20) becomes

$$H = g\mu_B \sum_i \mathbf{B}\hat{\mathbf{J}}_i - \frac{1}{2}\sum_{i,j} V(\mathbf{r}_i - \mathbf{r}_j)\hat{\mathbf{S}}_i\hat{\mathbf{S}}_j, \tag{6.28}$$

and it looks differently as compared to (6.1) in that the Zeeman term depends on a total momentum while the exchange interaction depends on spin operators only. If direct exchange interaction in Eq. (6.28) is induced by *s*-electrons the spin operators can be replaced with

$$\hat{\mathbf{S}} = (g - 1)\hat{\mathbf{J}} \tag{6.29}$$

This relation stems from Eqs. (6.22) and was introduced by De Gennes. If spin interaction is caused by electrons with finite orbital momentum, the relation (6.29) becomes approximate. A more detailed discussion of the approximate nature of Eq. (6.29) can be found in Ref. [10]. Using Eq. (6.28), the total Hamiltonian can be written as

$$H = g\mu_B \sum_{i} \mathbf{B}\hat{\mathbf{J}}_i - \frac{1}{2}(g-1)^2 \sum_{i,j} V(\mathbf{r}_i - \mathbf{r}_j)\,\hat{\mathbf{J}}_i\hat{\mathbf{J}}_j \tag{6.30}$$

The mean-field approach to the Hamiltonian (6.30) results in a Curie temperature similar to that which follows from Eq. (6.10):

$$T_c = \frac{1}{3}J(J+1)(g-1)^2 V_0. \tag{6.31}$$

The expression for magnetic moment (6.27) agrees well with various experimental measurements of the atomic magnetic moments of transition and rare earth elements embedded into a III-V host. Some illustrative examples are given below.

In III-V semiconductors, rare earth ions RE^{3+} substitute cation atoms. Outer electronic shells of isolated *Er* atom have configuration $4f^{12}5d^{0}6s^{2}$. In a GaN lattice two *s*-electrons and one *f*-electron go to the chemical bonds, so the configuration of the outer electronic shell becomes $4f^{11}$. The outer shells of the Tb atom have the configuration $4f^{9}5d^{0}6s^{2}$. After three electrons are transferred to the lattice the configuration of Tb^{3+} becomes $4f^{8}$. Shell $4f$ is weakly disturbed by the lattice because the shell is deep and screened from the crystal field by the outer *s*-shell.

Ion Er^{3+}. The electrons in the *f*-shell carry orbital angular momentum $l = 3$. They are distributed among $2l + 1$ energy levels, each corresponding to eigenvalues of the operator L_z: $m = -l, -l + 1 \ldots l$. Eleven *f*-electrons in Er^{3+} are distributed among these levels as shown in Table 6.1.

Table 6.1 Structure of Er^{3+} *f*-shell

↑	↑	↑	↑↓	↑↓	↑↓	↑↓
−3	−2	−1	0	1	2	3

The upper row in Table 6.1 depicts electrons each of which carries spin of ½ (arrows). The lower row shows the corresponding quantum number *m*. In order to find the total angular momentum of the shell we calculate the total orbital momentum of the shell *L* and total spin momentum *S* separately, and then sum them up, $J = L + S$. This implies that orbital and spin momenta are well defined. In other words, the spin-orbit interaction is weak enough that the energy splitting in the magnetic field between levels with different *m* is larger than the spin-orbit splitting (fine structure). This limit is called Russell–Saunders- or *LS*-coupling. For heavy elements the spin-orbit interaction becomes large, *LS*-coupling is not applicable, and the *j*–*j* coupling defines the electron configuration.

Table 6.1 is an example of the basic rules of how to fill out the electronic shell. **Hund's first rule**: the electron configuration in a shell should provide maximum spin. The exchange part of Coulomb interaction between electrons in a shell can be represented by terms $-V\mathbf{s}_i\mathbf{s}_j$. Parallel spins minimizes the energy. That is why three electrons in Table 6.1 are placed to have different orbital momenta thus maximizing the total spin of the shell. Obeying Hund's first rule we have to follow the Pauli principle: filling the state with two electrons with opposite spins. **Hund's second rule**: obeying the first rule keep the orbital angular momentum at a maximum. The spin-orbit interaction $\alpha\mathbf{LS}$ splits energy levels and lowers the total electron energy if first and second rules hold. That is why the pairs of electrons in Table 6.1 are placed in the cells with maximum values of *m*. **Hund's third rule**: $J = |L - S|$ if the shell is less half-filled, and $J = |L + S|$ if the shell is more than half full.

The total angular momentum in the example of Table 6.1 is $J = |L + S|$. The total orbital momentum of the shell can be calculated by adding quantum numbers *m* multiplied by a corresponding number of electrons carrying this momentum. Total spin comprises three uncompensated electron spins each carrying spin 1/2:

$$L = -3 - 2 - 1 + 1 * 2 + 2 * 2 + 3 * 2 = 6;$$

$$S = \frac{3}{2}; J = L + S = 15/2. \quad (6.32)$$

For Er^{3+} from Eq. (6.26) and (6.27) we find $g = 1.2$; $M = g\mu_B\sqrt{J(J+1)} = 9.58\mu_B$.

Ion Tb^{3+}. A corresponding diagram for eight *f*-electrons in Tb^{3+} is shown in Table 6.2.

Table 6.2 Structure of Tb^{3+} *f*-shell

↑	↑	↑	↑	↑	↑	↑↓
−3	−2	−1	0	1	2	3

$$L = -3 - 2 - 1 + 1 + 2 + 2 * 3 = 3$$

$$S = 3\,; J = L + S = 6; g = 1.5; M = g\mu_B\sqrt{J(J+1)} = 9.72\,\mu_B \quad (6.33)$$

Isolated Gd atom. The outer shells of an isolated Gd atom are in the configuration $4f^7 5d^1 6s^2$. Two *s*-electrons do not contribute to spin and orbital momenta of the shell. The relevant configuration is shown in Table 6.3.

Table 6.3 Structure of *d*- and *f*-shells in an isolated *Gd* atom

0	0	0	0	↑
−2	−1	0	1	2

d-shell

↑	↑	↑	↑	↑	↑	↑
−3	−2	−1	0	1	2	3

f-shell

Total orbital and spin momenta are $L = 2$, $S = 4$. The filling factor of shells taking together, is less than 1/2, so the total angular momentum $J = |L - S| = 2$. As follows from Eqs. (6.26) and (6.27), $g = 2.67$, $M = 6.53\mu_B$. Ion Gd^{2+} ($4f^7 5d^1 6s^0$) has the same electron configuration in the *f*- and *d*-shells as an isolated atom, so it is characterized by the same values of *g*-factor and magnetic moment.

GaN(Gd^{3+}). Seven *f*-electrons are in the outer shell when an isolated atom is placed at a cation site in the GaN lattice. The calculated g-factor and magnetic moment are shown in Table 6.4 and Eq. (6.34):

Table 6.4 Structure of Gd^{3+} *f*-shell

↑	↑	↑	↑	↑	↑	↑
−3	−2	−1	0	1	2	3

$$L = -3 - 2 - 1 + 1 + 2 + 3 = 0,$$

$$J = S = \frac{7}{2}; g = 2; M = g\mu_B\sqrt{J(J+1)} = 7.94\,\mu_B; \qquad (6.34)$$

Calculated g-factors and magnetic moments for the 4*f*-ions, Eqs. (6.32), (6.33), and (6.34) are in good agreement with experimental values.

GaAs(Mn)-GaN(Mn). Mn impurities ($3d^5 4s^2$) substitute for Ga ($3d^5 4s^2 p^1$), so the two *s*-electrons from Mn contribute to the crystal bonding the same way as Ga *s*-electrons do. The atomic structure of a Ga atom includes one *p*-electron which is absent in Mn, so Mn ions act as acceptors which can be in either neutral (Mn^{3+}) or charged (Mn^{2+}) states. Tetrahedral and hexagonal (in wurtzite GaN) crystal fields split the 3*d*-level in triply degenerate (for each spin) t_2- and doubly degenerate *e*-multiplets. In the ground state a hole is bound to a neutral Mn^{2+}acceptor (Mn^{2+} + hole) while converting into a charged Mn^{3+} when the chemical potential crosses the Mn^{3+}/Mn^{2+} level in the course of gradually increasing *n*-type co-doping [11].

As the crystal-field splitting in GaAs and GaN is smaller than the repulsion of two electrons in the state of the same momentum, the electron distribution in the Mn *d*-shell preserves high spin configuration. The angular momentum of Mn^{2+} in a high-spin configuration stems from *d*-electrons only: $J = S = 5/2$, $g_1 = 2$, $M_1 = g_1\,\mu_B\sqrt{J(J+1)} \approx 5.92\,\mu_B$. The total angular momenta of the complex Mn^{2+} + hole comprises momenta of Mn and a hole coupled antiparallel through the *p*-*d* coupling. The total angular momentum of the complex is $F = |S - j| = 1$, $J = s + l = 3/2$, where $l = 1$, $s = 1/2$ are the orbital momentum of the *p*-state at the valence band edge and the spin of the hole, respectively [12]. The complex as a whole can be described by the electron configuration $S = |S - s| = 2$, $L = 1$, $F = |L - S|$ which results in g-factor $g_2 = 2.5$ and magnetic moment $M_2 = g_2 \mu_B \sqrt{F(F+1)} \approx 4.33\,\mu_B$. This approximate estimation of the magnetic moment is in agreement with the experimental value in GaAs(Mn) reported in Ref. [13].

So, the material may contain Mn in the valence state fluctuating between Mn^{3+} and Mn^{2+} configurations. The main difference between hosts GaN and GaAs is the position of magnetic atom levels with respect to the conduction and valence band edges. In GaN, one of the localized states that originates from the split by a crystal field Mn *d*-level lies close to the valence band maximum. If the position of the Fermi level is close to the localized *d*-state it makes the valence of the impurity fluctuate and affects ferromagnetic phase transition. Ferromagnetism related to the fluctuating valence of the impurity in a metal was considered long time ago in Ref. [14].

In the next section, we describe the ferromagnetic phase transition in mixed valence III-V DMS.

6.4 Ferromagnetic Transition in a Mixed Valence Magnetic Semiconductor

Manganese atoms that substitute Ga in GaN and GaAs have a *d*-shell in neutral Mn^{3+} and charged Mn^{2+} states. Tetrahedral crystal field splits the *d*-level, so part of the levels fall deep into the conduction and valence bands far from the Fermi level. This makes them irrelevant in conduction and magnetism. Two levels fall in the GaN bandgap at 1.8 eV and 0.37 eV above the valence band edge [11, 15]. In GaAs the Mn level lies less than 0.1 eV above the valence band maximum so that an impurity band at high Mn content, merges with the valence band [8, 16]. If the chemical potential is close to the Mn^{3+} impurity level the impurities in different charge and magnetic states are coexistent in the sample. The simultaneous presence of Mn^{2+} ions in GaN has been confirmed experimentally [17].

Impurity levels associated with the Gd *d*-level are spin-split by *p*(*s*)-*f* exchange interaction with the band states giving rise to the states deep inside the valence and conduction bands [18], while the *d*-shell creates an impurity band in the GaN bandgap near the valence band [19] making localized spin a subject of mixed valence transition of $Gd^{3+/2+}$ in p-type GaN(Gd) samples.

6.4.1 Hamiltonian and Mean-Field Approximation

Non-magnetic co-doping determines the Fermi level position. When the Fermi level moves across the impurity level or impurity band of finite width, it changes the magnetic state of the impurity as the angular, spin, and total momentum are different in two charged states. For example, the parameters of Mn and Gd in two charge states are: Mn^{3+} ($J_1 = 1, S_1 = 2, g_1 = 2.5$); Mn^{2+} ($J_2 = S_2 = 5/2, g_2 = 2$); Gd^{3+} ($J_1 = S_1 = 7/2, g_1 = 2$); Gd^{2+} ($J_2 = 2, S_2 = 4, g_2 = 2.67$).

The magnetic field-induced spin splitting along with the charge states of an impurity can be taken into account using the Hamiltonian given below:

$$H = \mu_B g_0 B_0 \hat{J}_0^z (1 - \hat{n}) + \left[\varepsilon + \mu_B g_1 B_1 \hat{J}_1^z \right] \hat{n}, \tag{6.35}$$

where $\hat{n} = a^+ a$, a^+ is the electron creation operator in the d-orbital of energy E_1 inside the bandgap, $\varepsilon = E_1 - \mu$, μ is the chemical potential, g_n and $\hat{J}_n^z$ are the g-factors and the components of a d-orbital total angular momentum in the direction of the effective magnetic field, respectively, $n = 1, 0$ are the eigenvalues of $\hat{n}$. The effective magnetic field acting on impurity spin follows from the exchange interaction in Eq. (6.30):

$$B_n = B_z + \frac{(g_n - 1)^2 V_0 m x}{(g_n \mu_B)^2}, \tag{6.36}$$

where m is the average magnetic moment, $x = N_m/N_0$ is the number of magnetic ions related to the number of cation sites.

The position of the chemical potential μ is determined by Ga-vacancies and other non-magnetic defects. So, non-magnetic co-doping can be described by the variable ε which is the Fermi level position relative to the impurity level. Further in the text we use the term Fermi level despite the fact that μ may be in the bandgap. Co-doping of n-type corresponds to $\varepsilon < 0$: the Fermi level is higher in energy than the d-level making the level (or impurity band) completely filled at $T = 0$.

Model Eq. (6.35) describes the d-orbital in two charge states, filled ($n = 1$) and empty ($n = 0$), each of them $(2J_n + 1)$–fold degenerate. The thermodynamic characteristics of ferromagnetism can be found using the mean-field model (6.35) and the partition function (6.5). The trace operation in Eq. (6.5) should be performed taking into account that the dimension of the d-orbital multiplet as well as g-factors depend on the charge state. For example, $n = 1$ corresponds to Mn^{2+} whereas the state $n = 0$ corresponds to Mn^{3+}. Then the d-orbital partition function, electron distribution function $f = \langle \hat{n} \rangle$, and the magnetic moment under an external magnetic field, are given as

$$Z = Tr[\exp(-\beta \mathrm{H})] = \frac{\sinh\left[\left(J_1 + \frac{1}{2}\right)x_1\right]\exp(-\varepsilon/T)}{\sinh(x_1/2)} + \frac{\sinh\left[\left(J_0 + \frac{1}{2}\right)x_0\right]}{\sinh(x_0/2)},$$

$$f = T\frac{\partial(\ln Z)}{\partial \mu} = \frac{\sinh\left[\left(J_1 + \frac{1}{2}\right)x_1\right]}{Z\sinh(x_1/2)}\exp(-\varepsilon/T),$$

$$m(B_z) = T\frac{\partial(\ln Z)}{\partial B_z} = \frac{\mu_B}{Z}[g_1 G_1 \exp(-\varepsilon/T) + g_0 G_0], \tag{6.37}$$

where

$$G_n = \left\{\left(J_n + \frac{1}{2}\right)\cosh\left[\left(J_n + \frac{1}{2}\right)x_n\right] - \frac{1}{2}\coth\left(\frac{x_n}{2}\right)\sinh\left[\left(J_n + \frac{1}{2}\right)x_n\right]\right\}\Big/\sinh\left(\frac{x_n}{2}\right),$$

$$x_n = \mu_B g_n B_n/T. \tag{6.38}$$

When the effective magnetic field tends to zero, $B_n \to 0$, the filling factor becomes the Fermi distribution function of a localized level with different degrees of degeneracy in empty and filled states:

$$f = \frac{2J_1 + 1}{2J_1 + 1 + (2J_0 + 1)\exp(-\varepsilon/T)} \tag{6.39}$$

In completely empty or filled states of d-level the function $G_n(x)/Z$ tends to the Brillouin function $B_{J_n}(x)$.

At low temperature, $T \ll |\varepsilon|$, the d-orbital magnetic moment in Eq. (6.37) has different limits in filled and empty states. Accordingly, $m(B_z) \to \mu_B g_n J_n B_{J_n}$, where $n = 0$ if $\varepsilon > 0$ and $n = 1$ if $\varepsilon < 0$.

The mean-field critical temperature was discussed in detail in Section 6.1. The equation for T_c follows from the condition, $\partial m(B_z)/\partial m|_{B_z \to 0, m \to 0}$

$$T_c = \frac{1}{3}V_0 x F(T_c),$$

$$F(T_c) \equiv \frac{(g_1 - 1)^2 J_1(2J_1 + 1)(J_1 + 1)\exp(-\varepsilon/T_c) + (g_0 - 1)^2 J_0(2J_0 + 1)(J_0 + 1)}{(2J_1 + 1)\exp(-\varepsilon/T_c) + 2J_0 + 1} \tag{6.40}$$

In non-degenerate semiconductors the role of p- or n- non-magnetic co-doping in T_C is represented by the ε-dependent terms in Eq. (6.40): co-doping of $n(p)$-type corresponds to a negative (positive) ε. It is easy to find the relation between T_C that follows from Eq. (6.40) and the mean-field critical temperature (6.31). In the limit $\varepsilon >> T_C$ the position of the Fermi level μ is far below the d-level so that the level is empty and the limiting value of $F(T_C)$ in Eq. (6.40) gives

$$T_c = \frac{1}{3}(g_0 - 1)^2 V_0 x. \tag{6.41}$$

In the opposite limit, the level is filled, $|\varepsilon| >> T_C$, and the critical temperature tends to

$$T_c = \frac{1}{3}(g_1 - 1)^2 V_0 x. \tag{6.42}$$

So, Eq. (6.40) generalizes the mean-field approximation to account for the mixed valence of impurities taking part in ferromagnetic interaction.

To further analyze T_C one has to make assumptions on the interaction constant V_0. In transition metal or rare earth-doped GaN the free-carrier-induced indirect exchange interaction is questionable as as in most experimental situations typical of wide bandgap semiconductors, the electrons and holes are non-degenerate, so Fermi energy does not exist. Possible mechanisms of the interspin interaction rely on the excitation transfer through the magnetic impurity band, that is the double-exchange mechanism, or valence-impurity band transitions (see Chapter 5).

Let us consider the double-exchange mechanism first. As the mechanism relies on a carrier transfer within the impurity band, it is inefficient if the band is either totally filled or empty. This can be taken into account assuming that the magnetic coupling depends on a filling factor: $V_0(T) = V_0 f(1 - f)$. This results in the solution to Eq. (6.40) that exists in a finite range of ε determined by the thermal width of the Fermi filling factor as shown in Fig. 6.4. The actual range within which the d-level is magnetically active is also determined by the width of the impurity band.

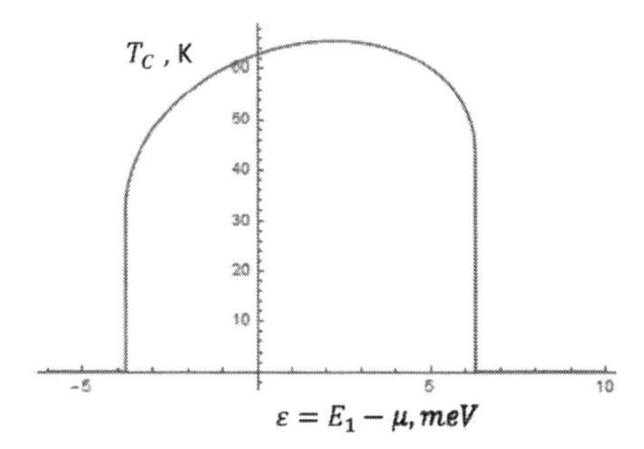

Figure 6.4 Double-exchange mean-field critical temperature, V_0 = 0.5 eV, x = 0.02.

As seen from Fig. 6.4, T_c is sensitive to a non-magnetic doping: as the Fermi level moves across the bandgap the impurity band width and fill-factor both determine the range of co-doping where magnetic impurities induce ferromagnetism.

Another interaction mechanism that does not rely on the free-carrier indirect exchange of RKKY-type is that based on impurity band-valence (or conduction) band transitions (IB-VB model, see Chapter 5). This mechanism works when the Fermi level is located between the valence band and the acceptor impurity band no matter whether the impurity band is empty or not. So, interspin interaction becomes ineffective only if the Fermi level goes up deeply in the bandgap making the impurity band totally filled up: $V_0(T) = V_0\,(1 - f)$. The solution to Eq. (6.40) is shown in Fig. 6.5.

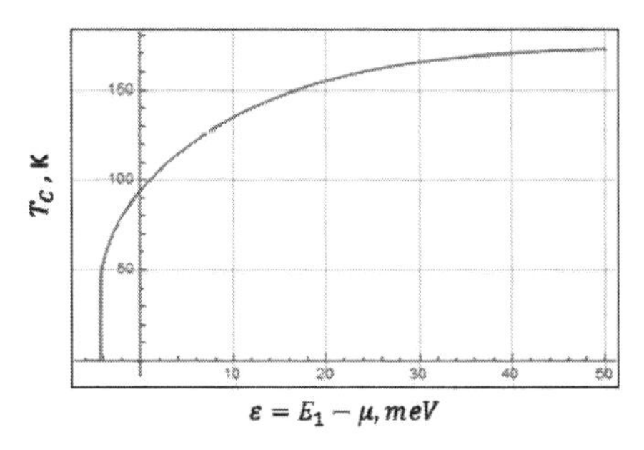

Figure 6.5 Mean-field critical temperature within the IB-VB model V_0 = 0.5 eV, x = 0.02.

The absolute value of the critical temperature depends on parameter V_0 which is not exactly known in GaN.

6.4.2 Percolation

In a low carrier density regime, which is most likely the case in a wide bandgap semiconductor, the indirect exchange interaction is short-ranged and the mean-field approximation cannot be applied. The onset of ferromagnetism occurs due to the formation of ferromagnetic clusters that grow and merge until a single cluster penetrates the whole crystal. Unlike that in the mean field approximation, the percolation critical temperature is sensitive to the shape of the range function. Assuming that the mechanism of pair exchange interaction is attributed to the impurity band-valence band transitions, the range function can be written as

$$V(R) = -I(1-f)\left\{\left[\frac{d_0}{R+d_0}\right]^{5/2} \exp(-R/r_0) - b\left(\frac{d_0}{R+d_0}\right)^{10}\right\}, \quad (6.43)$$

where I is the coupling coefficient, the second term in $V(R)$ is the antiferromagnetic coupling between closed pairs [20], b is the fitting parameter. The interaction radius is determined by the impurity level activation energy as follows from Eq. (5.44), $r_0 = \hbar(2m_v E_1)^{-1/2}$, where E_1 is referenced to a valence band maximum. Interaction (6.43) is illustrated in Fig. 6.6.

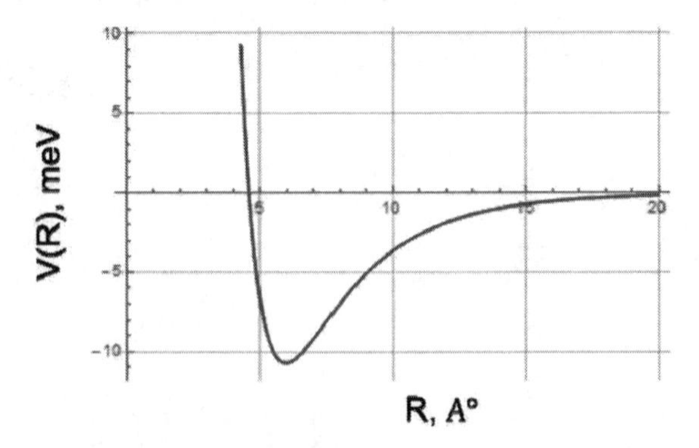

Figure 6.6 Indirect exchange interaction truncated by antiferromagnetic coupling at short distances, $b = 80$, $E_1 = 0.35$ eV, $d_0 = 4.5$ Å.

A pair of spins at a distance R aligns parallel if the temperature becomes as low as $T = \frac{1}{3}V(R)F(T)$. When the temperature goes down toward the critical value, the ferromagnetic clusters grow, and the critical point is reached when the cluster size equals to the percolation radius which is proportional to the average interspin distance, $r_c = \sqrt[3]{2.4}\,\overline{R}$ (see details in Section 6.2):

$$T_c = \frac{1}{3} V(r_c) F(T_c) \tag{6.44}$$

The numerical solution to Eq. (6.44) is illustrated in Fig. 6.7.

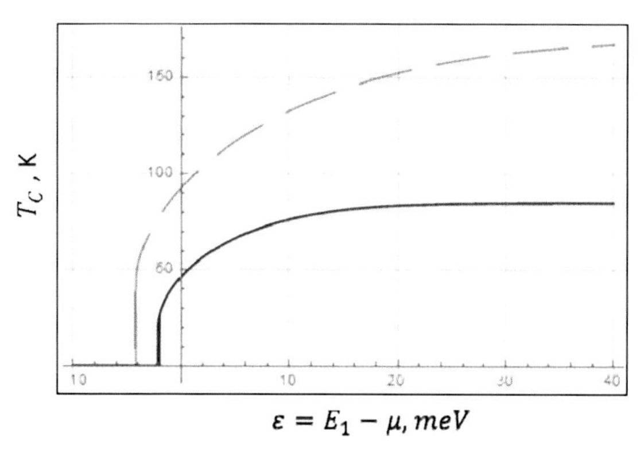

Figure 6.7 Percolation critical temperature as a function of the Fermi level position (non-magnetic co-doping) at different Mn-content. Solid line: $x = 0.02$; dashed line: $x = 0.05$.

The critical temperature increases when p-type doping shifts the Fermi level closer to a valence band, $\varepsilon > 0$. Figure 6.8 shows the critical temperature versus the Mn-content.

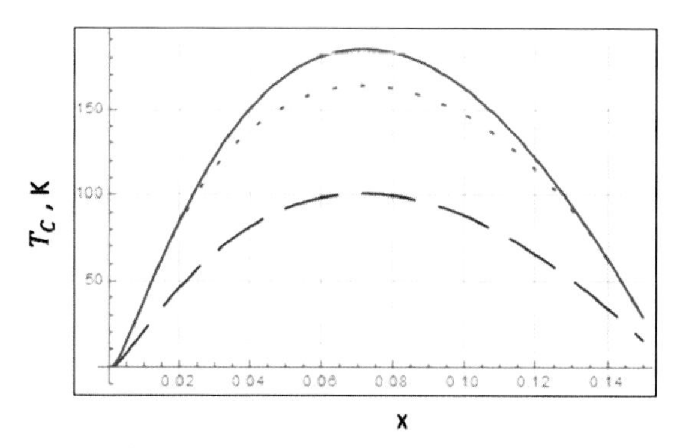

Figure 6.8 Plot of the percolation critical temperature as a function of a transition metal content at various co-doping levels. Solid line: $\varepsilon = 60$ meV dotted line: $\varepsilon = 20$ meV; dashed line: $\varepsilon = 0$.

When x increases the linear x-dependence of T_c for small x changes to a bell-shape because higher Mn-content implies a shorter average interspin distance for which the interaction sign favors antiferromagnetic coupling and suppresses ferromagnetism. The critical temperature is sensitive to the shape of the range function. Results shown in Fig. 6.8, are in a good agreement with those obtained from first-principle calculations [9].

6.5 Conclusions

The MFA and percolation models of ferromagnetism are extended to account for the mixed valence and mixed magnetic state of a transition metal impurity in III-nitride semiconductors. The ferromagnetic critical temperature, calculated using the mean-field and percolation scenario of the phase transition, strongly depends on the Fermi level position relative to the TM level in the bandgap. The results of percolation theory qualitatively agree with those obtained from the first-principle calculations.

To date, the mechanism of indirect exchange interaction and the existence of room-temperature ferromagnetism are the subject of discussions presenting most controversial issues in the physics of wide bangap III-V dilute magnetic semiconductors. Room-temperature ferromagnetism in III-nitrides has been confirmed in various independent experiments. However, the origin of ferromagnetism has not yet been established. Whether it is attributed to magnetic clusters or exists in homogeneously magnetically doped materials remains unclear. The absolute value of the critical temperature obtained with microscopic models in this chapter relies on parameters to be determined from comparison between theory and experiment.

Problems

6.1 Find the cluster size and its temperature dependence if the range function has the form:

$V(R) = \beta \exp(-\alpha R)$.

6.2 Calculate the magnetic moment of ion Dy^{3+} with electron configuration $4f^9$.

6.3 Show that the magnetic moment of Eu^{3+} is equal to zero.

References

1. Tanaka M, Ohya S, Hai PN (2014) Recent progress in III-V based ferromagnetic semiconductors: Band structure, Fermi level, and tunneling transport, *Appl Phys Rev*, **1**, 011102.
2. Dietl T, Ohno H, Matsukura F, Gibert J, Ferrand D (2000), Zener model description of ferromagnetism in zinc-blende magnetic semiconductors, *Science*, **287**, 1019–1022.
3. Jungwirth T, Sinova J, Kusera J, MacDonald AH (2006) Theory of ferromagnetic (III,Mn)V semiconductors, *Rev Mod Phys*, **78**, 809–864.
4. Matsukura F, Ohno H, Shen A, Sugawara Y (1998) Transport properties and origin of ferromagnetism in (Ga, Mn)As, *Phys Rev B*, **57**, R2037; Ohno H, Matsukura F (2001) A ferromagnetic III–V semiconductor: (Ga, Mn)As, *Solid State Commun*, **117**, 179.
5. Korenblit IYA, Shender EF (1978) *Sov Phys Usp*, **21**, 832; Shklovskii BI, Efros AL (1984) *Electronic Properties of Doped Semiconductors*, Springer, New York.
6. Litvinov VI, Dugaev VK (2001) Ferromagnetism in magnetically doped III-V semiconductors, *Phys Rev Lett*, **86**(24), 5593–5596; Litvinov VI (2005) Mixed valence, percolation, and ferrromagnetism in transition metal-doped GaN, *Phys Rev B*, **72**, 195209; Litvinov VI, Dugaev VK (2009) Room-temperature ferromagnetism in dielectric GaN(Gd), *Appl Phys Lett*, **94**, 212506.
7. Chapler BC, Mack S, Myers RC, Frenzel A, Pursley BC, Burch KS, Dattelbaum AM, Samarth N, Awschalom DD, Basov DN (2013), Ferromagnetism and infrared electrodynamics of $Ga_{1-x}Mn_xAs$, *Phys Rev B*, **87**, 205314.
8. Schilfgaarde M, Mryasov ON (2001) Anomalous exchange interactions in III-V dilute magnetic semiconductors, *Phys Rev B*, **63**, 233205; Kulatov E, Nakayama H, Mariette M, Ohta H, Uspenskii Yu A (2002) Electronic structure, magnetic ordering, and optical properties of GaN and GaAs doped with Mn, *Phys Rev B*, **66**, 0452203.
9. Sato K, Katayama-Yoshida H (2002) First principles materials design for semiconductor spintronics, *Semicond Sci Technol*, **17**, 367–376; Sato K, Dederics PH, Katayama-Yoshida H (2003) Curie temperatures of III–V diluted magnetic semiconductors calculated from first principles, *Europhys Lett*, **61**(N3), 403; Dinh VA, Sato K, Katayama-Yoshida H (2005) Curie temperatures of cubic (Ga, Mn) N diluted magnetic semiconductors from the RKKY spin model, *J Supercond*, **18**, 47–53.

10. Irkhin Yu P (1988) Electron structure of the 4f shells and magnetism of rare-earth metals, *Sov Phys Usp*, **31,** 163–170.
11. Graf T, Gjukic M, Brandt MS, Stutzmann M (2002) The Mn acceptor level in group III nitrides, *Appl Phys Lett*, **81**(27), 5159–5161.
12. Sapega VF, Ruf T, Cardona M (2000) Spin-flip Raman scattering in Mn-doped GaAs: Exchange interaction and *g*-factor renormalization, *Solid State Commun*, **114**, 573–577.
13. Jungwirth T, Mašek J, Wang KY, Edmonds KW, Sawicki M, Polini M, Sinova J, MacDonald AH, Campion RP, Zhao LX, S.Farley NR, Johal TK, van der Laan G, Foxon CT, Gallagher BL (2006) Low-temperature magnetization of {(Ga,Mn)As} semiconductors, *Phys Rev B*, **73**, 165205; preprint arXiv:cond-mat/0508255.
14. Abrikosov AA (1974) On the theory of impurity ferromagnetism in semiconductors, *Sov Phys JETP*, **38**(2), 403–407.
15. Barthel S, Kunert G, Gartner M, Stoica M, Mourad D, Kruse C, Figge S, Hommel D, Czycholl G (2014) Determination of the Fermi level position in dilute magnetic GaMnN films, *J Appl Phys*, **115**, 123706.
16. Muneta I, Terada H, Ohya S, Tanaka M (2013) Anomalous Fermi level behavior in GaMnAs at the onset of ferromagnetism, *Appl Phys Lett*, **103**, 032411.
17. Sonoda S, Tanaka I, Ikeno H, Yamamoto T, Oba F, Araki T, Yamamoto Y, Suga K, Nanishi Y, Akasaka Y, Kindo K, Hori H (2006) Coexistence of Mn^{2+} and Mn^{3+} in ferromagnetic GaMnN, *J Phys Condens Matter* **18**, 4615–4621.
18. Zhong GH, Wang JL, Zeng Z (2008) Electronic and magnetic structures of 4f in $Ga_{1-x}Gd_xN$, *J Phys Condens Matter*, **20**, 295221.
19. Liu L, Yu PY, Ma Z, Mao SS (2008) Ferromagnetism in GaN:Gd: A density functional theory study, *Phys Rev Lett*, **100**, 127203.
20. Inoue J (2003) Effective exchange interaction and Curie temperature in magnetic semiconductors, *Phys Rev B*, **67**, 125302.

Chapter 7

Topological Insulators

Various materials and nanostructures with large Rashba coefficient are a subject of study in spintronic material science as they present templates for voltage-controlled spintronic applications. Some of these material systems were mentioned in Chapter 3. There is a specific class of semiconductors called topological insulators (TI) that presents new opportunities to spintronics. TI surface electron states experience a strong spin-orbit interaction and can be manipulated by an external electric field. A large linear-k Rashba coefficient has been reported in Bi_2Te_3-based TI [1].

A topological insulator is a semiconductor with a specific band structure that makes the bulk electrically insulating and the surfaces—conducting. More specifically, the electron spectrum has an energy gap in the bulk (semiconductor) and it is gapless at the surfaces (metal). If the surface electron spectrum comprises odd number of minima in the Brillouin zone, the minima are gapless (Dirac cones), they are protected by time reversal symmetry, and lie in the bulk energy gap. Perturbations which do not violate the symmetry cannot open the gap at the surface and destroy the conduction state [2–4]. The position of the TI surface branches of the spectrum relative to the bulk bandgap is illustrated in Fig. 7.1.

The linear dispersion of gapless surface excitations, shown in Fig. 7.1, forms a Dirac cone in variables E, k_x, k_y. Due to the spin-orbit interaction, the electron momentum is locked to the spin

Wide Bandgap Semiconductor Spintronics
Vladimir Litvinov

ISBN 978-981-4669-70-2 (Hardcover), 978-981-4669-71-9 (eBook)
www.panstanford.com

orientation so that electrons with opposite spins move in opposite directions. This makes non-magnetic scattering on the surface ineffective as the change of momentum is associated with the spin rotation and thus the scattering is forbidden unless the localized impurity spin takes part in the scattering process preserving the total (electron plus impurity) spin conservation. Non-dissipative edge currents are illustrated in Fig. 7.2, where a two-dimensional TI is shown as an example.

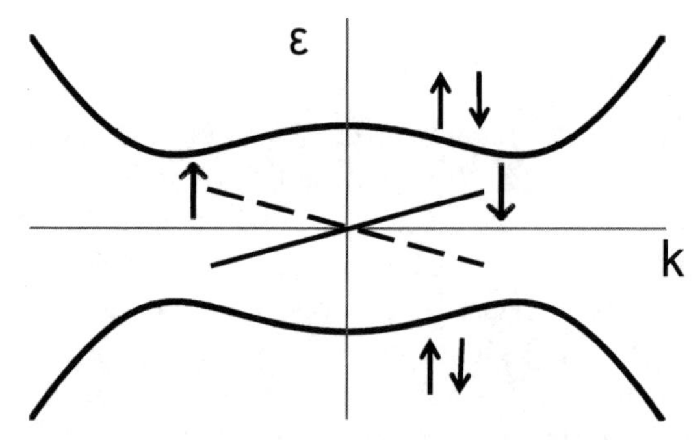

Figure 7.1 Linear in-plane electron energy spectrum on a single surface. Branches ↑↓ correspond to up- and down-spin electrons. Bold solid lines: bulk energy dispersion diagram.

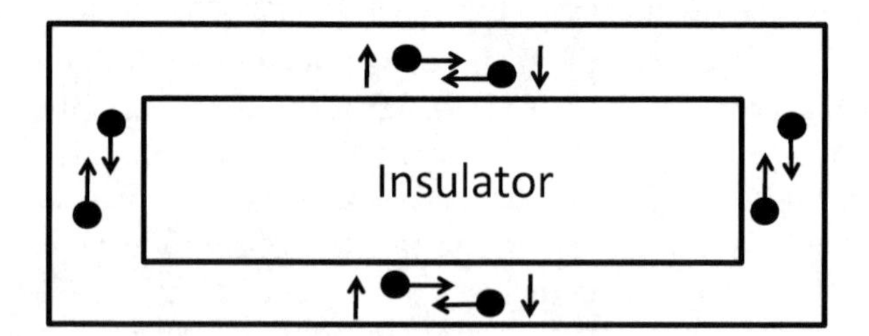

Figure 7.2 Non-dissipative edge currents in topological insulator.

In equilibrium, electron fluxes along the edge compensate each other and no electric charge transfer takes place. If voltage is applied along the, the electric field induces a net electric current that makes non-equal the number of electrons moving in opposite directions. The imbalance, illustrated in Fig. 7.3 by different Fermi momenta of ↑↓ electrons, translates to a spin accumulation in the direction perpendicular to an electric current. That is what is called the quantum spin Hall effect.

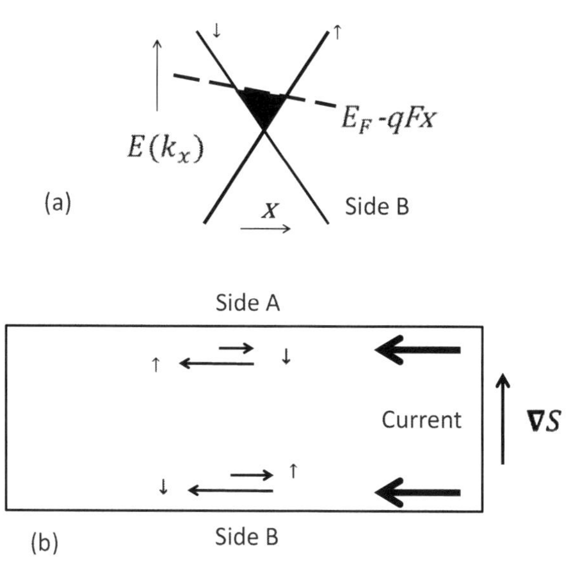

Figure 7.3 Quantum spin Hall effect. (a) Occupied electron states (black area) in imbalanced surface electron system, (b) net electric current and spin flow in a biased surface.

As the TI phase can be characterized by the quantum spin Hall effect, it is often called the spin Hall phase. Experimental evidence of two spin-polarized counter-propagating edge states is reported in Ref. [5].

The TI state has more often been experimentally observed in $Bi_2(Te,Se)_3$ [6, 7] and HgTe/CdHgTe [3] material systems with the inverted bulk band spectra. Generally, the time reversal symmetry, the spin-orbit interaction, and the inverted band spectrum are the necessary conditions for a TI state to exist in a semiconductor. One more material that shows TI properties is an InGaN/GaN quantum well wide enough to provide an inverted position of the conduction (Γ_1) and light hole (Γ_6) bands [8].

7.1 Bulk Electrons in Bi_2Te_3

In order to differentiate trivial and topological phases it is instructive to look at the bulk energy spectrum of the material. Relevant electron states have energy close to the chemical potential and stem from linear combinations of **Bi** p-orbitals $|p1_z^+, s\rangle$ and linear combinations of **Te** (or **Se**) p-orbitals $|p2^{\pm}_{z}, s\rangle$, where s is the spin index, and ± labels the parity of the linear combinations. Until the spin-orbit interaction is not taken into account, the conduction and valence band edges are in the normal position corresponding to even (+) and odd (−) functions, respectively. Spin-orbit interaction shifts the energy levels, resulting in inverted band edge positions, as shown in Fig. 7.4.

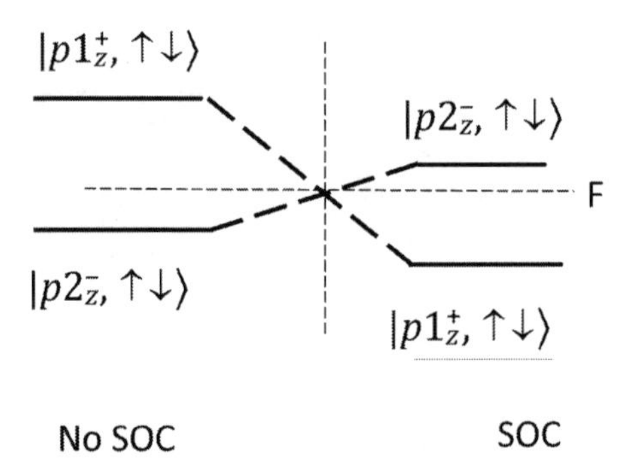

Figure 7.4 Band inversion caused by spin-orbit interaction (SOC).

Bloch amplitudes, $|v\uparrow\rangle = |p1^{+}_{z}, \uparrow\rangle$, $|c\uparrow\rangle = |p2^{-}_{z}, \uparrow\rangle$, $|v\downarrow\rangle = |p1^{+}_{z}, \downarrow\rangle$, $|c\downarrow\rangle = |p2^{-}_{z}, \downarrow\rangle$ create a basis which can be represented by a four-component function $\phi = \begin{pmatrix} |v\uparrow> \\ |c\uparrow> \\ |v\downarrow> \\ |c\downarrow> \end{pmatrix}$, so that the Schrödinger equation is written as $\boldsymbol{H}\phi = \boldsymbol{E}\phi$ where the 4 × 4 matrix Hamiltonian has the form

$$H = (D_1 k_z^2 + D_2 k^2) I + \begin{pmatrix} M(\boldsymbol{k}) + V(z) & A_1 k_z & 0 & A_2 k_- \\ A_1 k_z & -M(\boldsymbol{k}) + V(z) & A_2 k_- & 0 \\ 0 & A_2 k_+ & M(\boldsymbol{k}) + V(z) & -A_1 k_z \\ A_2 k_+ & 0 & -A_1 k_z & -M(\boldsymbol{k}) + V(z) \end{pmatrix} \quad (7.1)$$

where I = diag(1,1,1,1), $V(z)$ is the electron energy in an electric field that includes the external field and the possible internal near-surface field, $M(\boldsymbol{k}) = -\Delta - B_1 k_z^2 - B_2 k^2$, $k^2 = k_x^2 + k_y^2$, $k_{\pm} = k_x \pm ik_y$.

Numerical values of parameters, $A_{1,2}$, $B_{1,2}$ and $D_{1,2}$ can be found in Ref. [6].

In order to foresee the topological properties of the surface perpendicular to the z-axis one could inspect the bulk energy dispersion in the z-direction. As the constant B_1 is positive, the band dispersion in the z-direction has distinctive features depending on the sign of the energy gap 2Δ. Eigenvalues of the Hamiltonian (7.1) present the spin degenerate conduction and valence bands given below and illustrated in Fig. 7.5.

$$E_{C\uparrow\downarrow}(k) = D_1k_z^2 + D_2k^2 + \sqrt{A_1^2k_z^2 + A_2^2k^2 + (B_2k^2 + B_1k_z^2 + \Delta)^2},$$
$$E_{V\uparrow\downarrow}(k) = D_1k_z^2 + D_2k^2 - \sqrt{A_1^2k_z^2 + A_z^2k^2 + (B_2k^2 + B_1k_z^2 + \Delta)^2}. \quad (7.2)$$

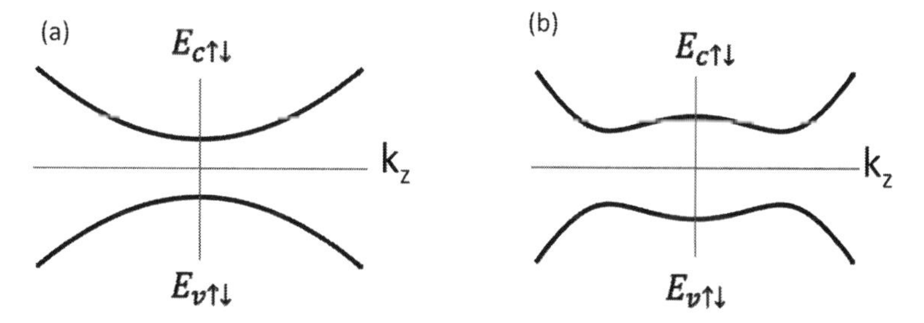

Figure 7.5 Energy dispersion in the z-direction. (a) Normal band positions, $\Delta > 0$; (b) inverted bands, $\Delta < 0$.

The spin-orbit induced twist in the bulk bands at $B_1\Delta < 0$ is what results in the existence of spin-polarized surface electrons with linear spectrum. In the next section we discuss in more detail how surface electrons acquire linear dispersion when the bulk spectrum is characterized by inverted level positions.

It should be noted that there are other types of topological insulators where inverted bands have nothing to do with the spin-orbit interaction and the topological phase reveals itself by the surface metallic states with massless electrons. These states are protected not by time reversal symmetry but rather by point crystalline symmetry. This class is termed a crystalline topological insulator and includes the $Pb_{1-x}Sn_xTe$ material system. A topological state occurs when Sn content exceeds a critical value, making transition from normal to inverted band positions [9–11].

7.2 Surface Dirac Electrons

Below we use the Hamiltonian (7.1) at $D_1 = D_2 = 0$ as this simplification does not prevent us from studying topological surface states. In the following discussion,we account for the finite size of the sample in the z-direction, making the substitution, $k_z \to -i\partial/\partial z$. Our goal is to obtain the surface energy spectrum at small k (in the vicinity of the Γ-point), so we will treat the finite-k part of the Hamiltonian H_1 as a perturbation:

$$H = H_0 + H_1,$$

$$H_0 = \begin{pmatrix} -\Delta + B_1\frac{\partial^2}{\partial z^2} & -iA_1\frac{\partial}{\partial z} & 0 & 0 \\ -iA_1\frac{\partial}{\partial z} & \Delta - B_1\frac{\partial^2}{\partial z^2} & 0 & 0 \\ 0 & 0 & -\Delta + B_1\frac{\partial^2}{\partial z^2} & iA_1\frac{\partial}{\partial z} \\ 0 & 0 & iA_1\frac{\partial}{\partial z} & \Delta - B_1\frac{\partial^2}{\partial z^2} \end{pmatrix},$$

$$H_1 = [V_S(z) + V_{AS}(z)]I + \begin{pmatrix} -B_2k^2 & 0 & 0 & A_2k_- \\ 0 & B_2k^2 & A_2k_- & 0 \\ 0 & A_2k_+ & -B_2k^2 & 0 \\ A_2k_+ & 0 & 0 & B_2k^2 \end{pmatrix}, \qquad (7.3)$$

where the external field $V(z)$ is written as the sum of $V_S(z) = \frac{1}{2}[V(z) + V(-z)]$ and $V_{AS}(z) = \frac{1}{2}[V(z) - V(-z)]$, even and odd parts, respectively. The Hamiltonian H_0 is a block-diagonal with upper and lower blocks corresponding to spin-up and down electrons, respectively. It is sufficient to consider the upper block only and then use the resulting wave functions substituting $A_1 \to -A_1$ to get the characteristics of the lower block. So, we will be dealing with a 2×2 Hamiltonian that describes spin-up electrons and holes.

$$H_0 = \begin{pmatrix} -\Delta + B_1\frac{\partial^2}{\partial z^2} & -iA_1\frac{\partial}{\partial z} \\ -iA_1\frac{\partial}{\partial z} & \Delta - B_1\frac{\partial^2}{\partial_z^2} \end{pmatrix}. \qquad (7.4)$$

To study surface states, we chose a wave function localized near the surface: $\psi_\uparrow = A\begin{pmatrix}\varphi_1 \\ \varphi_2\end{pmatrix}\exp(\lambda z)$. Using $\psi_\uparrow$ in the Schrödinger equation $H_0\psi_\uparrow = E\psi_\uparrow$ we get the characteristic equation for λ:

$$\det\begin{pmatrix} -\Delta + B_1\lambda^2 - E & -iA_1\lambda \\ -iA_1\lambda & -\Delta + B_1\lambda^2 - E \end{pmatrix} = 0. \tag{7.5}$$

The equation has four solutions: $\pm\lambda_1$, $\pm\lambda_2$,

$$\lambda_1 = \frac{1}{\sqrt{2}\,B_1}[F - \sqrt{F^2 + 4B_1^2(E^2 - \Delta^2)}]^{1/2},$$

$$\lambda_2 = \frac{1}{\sqrt{2}\,B_1}[F + \sqrt{F^2 + 4B_1^2(E^2 - \Delta^2)}]^{1/2},$$

$$F = A_1^2 + 2B_1\Delta. \tag{7.6}$$

To guarantee the exponential decay of the wave function away from the surface, we look for conditions under which λ_1, λ_2 are both real. Real λ_1, λ_2 exist only in the energy interval inside the energy gap $|E| < |\Delta|$. Outside the gap, the imaginery λ makes the wave function oscillatory and merging with the bulk states. For electron energy that falls within the gap, additional constrains following from Eq. (7.6) ensure the existence of well-defined surface states:

$$B_1\Delta < 0; A_1^2 > 4|B_1\Delta|. \tag{7.7}$$

So, surface states exist if the bulk energy spectrum is inverted. In order to find the energy spectrum of surface electrons we present the wave function $\begin{pmatrix}\varphi_1 \\ \varphi_2\end{pmatrix}$ as the superposition of two linearly independent eigenfunctions of the Hamiltonian (7.4) corresponding to both real λ_1 and λ_2:

$$\begin{pmatrix} -\Delta + B_1\lambda_1^2 + E \\ -iA_1\lambda_1 \end{pmatrix}; \begin{pmatrix} -\Delta + B_1\lambda_2^2 + E \\ -iA_1\lambda_2 \end{pmatrix} \tag{7.8}$$

As we consider a film with two surfaces at $z = \pm L/2$, the wave function also includes terms with $-\lambda_1$ and $-\lambda_2$:

$$\psi_{\uparrow}(z) = C_{11}\begin{pmatrix} B_1\lambda_1^2 + E - \Delta \\ -iA_1\lambda_1 \end{pmatrix}\exp(\lambda_1 z) + C_{12}\begin{pmatrix} B_1\lambda_1^2 + E - \Delta \\ iA_1\lambda_1 \end{pmatrix}\exp(-\lambda_1 z)$$
$$+ C_{21}\begin{pmatrix} B_1\lambda_2^2 + E - \Delta \\ -iA_1\lambda_2 \end{pmatrix}\exp(\lambda_2 z) + C_{22}\begin{pmatrix} B_1\lambda_2^2 + E - \Delta \\ iA_1\lambda_2 \end{pmatrix}\exp(-\lambda_2 z) \quad (7.9)$$

The problem has the symmetry with respect to spatial inversion: the surfaces at $z = \pm L/2$ are identical, so we can reduce the number of coefficients in Eq. (7.9) by assuming symmetric ($C_{11} = C_{12}$, $C_{21} = C_{22}$) and antisymmetric ($C_{11} = -C_{12}$, $C_{21} = -C_{22}$) superpositions in (7.9). Next we consider the two cases separately.

(1) $C_{11} = C_{12} = A$, $C_{21} = C_{22} = B$. The wave function takes the form

$$\chi_{\uparrow}(z) = A\begin{pmatrix} (B_1\lambda_1^2 + E - \Delta)\cosh[\lambda_1 z] \\ -iA_1\lambda_1 \sinh[\lambda_1 z] \end{pmatrix} + B\begin{pmatrix} (B_1\lambda_2^2 + E - \Delta)\cosh[\lambda_2 z] \\ -iA_1\lambda_2 \sinh[\lambda_2 z] \end{pmatrix}. \quad (7.10)$$

The zero boundary conditions $\chi_{\uparrow}(\pm L/2) = 0$ become a system of homogeneous equations for coefficients A, B

$$M\begin{pmatrix} A \\ B \end{pmatrix} = 0,$$

$$M = \begin{pmatrix} (B_1\lambda_1^2 + E - \Delta)\cosh[\lambda_1 L/2] & (B_1\lambda_2^2 + E - \Delta)\cosh[\lambda_2 L/2] \\ iA_1\lambda_1 \sinh[\lambda_1 L/2] & iA_1\lambda_2 \sinh[\lambda_2 L/2] \end{pmatrix}, \quad (7.11)$$

and the equation $\det[M] = 0$ determines the surface energy level E_-:

$$\frac{(B_1\lambda_1^2 + E - \Delta)\lambda_2}{(B_1\lambda_2^2 + E - \Delta)\lambda_1} = \frac{\tanh[\lambda_1 L/2]}{\tanh[\lambda_2 L/2]}. \quad (7.12)$$

Wave function corresponding to this energy level can be found from Eq. (7.10) providing the coefficients A, B are known. Using (7.11) and (7.12), one finds

$$B = -A\,\frac{(B_1\lambda_1^2 + E - \Delta)}{(B_1\lambda_2^2 + E - \Delta)} * \frac{\cosh[\lambda_1 L/2]}{\cosh[\lambda_2 L/2]}. \quad (7.13)$$

Solving (7.12) with respect to $E - \Delta$, we get

$$E - \Delta + B_1\lambda_1^2 = \frac{B_1\lambda_1(\lambda_1^2 - \lambda_2^2)\tanh[\lambda_1 L/2]}{\lambda_1\tanh[\lambda_1 L/2] - \lambda_2\tanh[\lambda_2 L/2]},$$

$$E - \Delta + B_1\lambda_2^2 = \frac{B_1\lambda_2(\lambda_2^2 - \lambda_1^2)\tanh[\lambda_2 L/2]}{\lambda_2\tanh[\lambda_2 L/2] - \lambda_1\tanh[\lambda_1 L/2]}. \quad (7.14)$$

Now substituting the coefficient B from (7.13) into Eq. (7.10) and using relations (7.14) we come to an explicit form of the wave function corresponding to energy level E_-:

$$\chi_\uparrow(z) = A\left[\begin{pmatrix} (B_1\lambda_1^2 + E - \Delta)\cosh[\lambda_1 z] \\ -iA_1\lambda_1\sinh[\lambda_1 z] \end{pmatrix} - \frac{(B_1\lambda_1^2 + E - \Delta)}{(B_1\lambda_2^2 + E - \Delta)} * \frac{\cosh[\lambda_1 L/2]}{\cosh[\lambda_2 L/2]} * \begin{pmatrix} (B_1\lambda_2^2 + E - \Delta)\cosh[\lambda_2 z] \\ -iA_1\lambda_2\sinh[\lambda_2 z] \end{pmatrix}\right] = C\begin{pmatrix} -B_1\eta_2(E_-)f_+(z, E_-) \\ iA_1 f_-(z, E_-) \end{pmatrix}, \quad (7.15)$$

where

$$\eta_2(E_-) = \left.\frac{\lambda_1^2 - \lambda_2^2}{\lambda_1\tanh[\lambda_1 L/2] - \lambda_2\tanh[\lambda_2 L/2]}\right|_{E_-},$$

$$f_+(z, E_-) = \left.\left\{\frac{\cosh[z\lambda_1]}{\cosh[\lambda_1 L/2]} - \frac{\cosh[z\lambda_2]}{\cosh[\lambda_2 L/2]}\right\}\right|_{E_-},$$

$$f_-(z, E_-) = \left.\left\{\frac{\sinh[z\lambda_1]}{\sinh[\lambda_1 L/2]} - \frac{\sinh[z\lambda_2]}{\sinh[\lambda_2 L/2]}\right\}\right|_{E_-}, \quad (7.16)$$

and C is the constant determined from the normalization condition $\int_{-L/2}^{L/2}\chi_\uparrow(z)^+\chi_\uparrow(z) = 1$.

(2) $C_{11} = -C_{12} = A$, $C_{21} = -C_{22} = B$. The wave function that follows from Eq. (7.9) is expressed as

$$\varphi_\uparrow(z) = A\begin{pmatrix} (B_1\lambda_1^2 + E - \Delta)\sinh[\lambda_1 z] \\ -iA\lambda_1\cosh[\lambda_1 z] \end{pmatrix} + B\begin{pmatrix} (B_1\lambda_1^2 + E - \Delta)\sinh[\lambda_2 z] \\ -iA_1\lambda_2\cosh[\lambda_2 z] \end{pmatrix}. \quad (7.17)$$

Boundary conditions $\varphi_\uparrow(\pm L/2) = 0$ result in an equation for the surface state E_+:

$$\frac{(B_1\lambda_1^2 + E - \Delta)\,\lambda_2}{(B_1\lambda_2^2 + E - \Delta)\lambda_1} = \frac{\tanh\,[\lambda_2 L/2]}{\tanh\,[\lambda_1 L/2]}. \tag{7.18}$$

Following the procedure described after Eq. (7.12) we obtain an explicit form of the wave function corresponding to the energy level E_+:

$$\varphi_\uparrow(z) = C\begin{pmatrix} -B_1\eta_1(E_+)f_-(z,E_+) \\ iA_1 f_+(z,E_+) \end{pmatrix}, \tag{7.19}$$

where

$$\eta_1(E_+) = \left.\frac{\lambda_1^2 - \lambda_2^2}{\lambda_1\coth\,[\lambda_1 L/2] - \lambda_2\coth\,[\lambda_2 L/2]}\right|_{E_+},$$

$$f_+(z,E_+) = \left.\left\{\frac{\cosh\,[z\lambda_1]}{\cosh\,[\lambda_1 L/2]} - \frac{\cosh\,[z\lambda_2]}{\cosh\,[\lambda_2 L/2]}\right\}\right|_{E_+},$$

$$f_-(z,E_+) = \left.\left\{\frac{\sinh\,[z\lambda_1]}{\sinh\,[\lambda_1 L/2]} - \frac{\sinh\,[z\lambda_2]}{\sinh\,[\lambda_2 L/2]}\right\}\right|_{E_+}, \tag{7.20}$$

Wave functions $\varphi_\uparrow(z)$ and $\chi_\uparrow(z)$ correspond to energy levels E_+ and E_-, respectively, and present surface electron states in Γ-point ($\mathbf{k} = 0$). The gap in the surface energy spectrum in Γ-point can be expressed analytically in the limit $\lambda_1 >> \lambda_2$, $L\lambda_{1,2} >> 1$:

$$\Delta_S = E_+ - E_- \approx 4\Delta\,\exp(-\lambda_2 L) \tag{7.21}$$

The spectrum becomes gapless if the film thickness increases so that the overlap of wave functions localized near opposite surfaces tends to zero.

Two blocks in the block-diagonal Hamiltonian H_0 corresponds to spin-up and spin-down electrons and they differ in the sign of the parameter A_1. So, wave functions $\varphi_\downarrow(z)$ and $\chi_\downarrow(z)$ can be obtained from spin-up functions $\varphi_\uparrow(z)$ and $\chi_\uparrow(z)$ by replacing $A_1 \to -A_1$. Energy levels $E_\pm$ do not change. This corresponds to the spin degeneracy of the spectrum. Four eigenspinors of the Hamiltonian H_0 describe two surface levels with two spins and can be written as

$$\Phi_1(z) = \begin{pmatrix} \varphi_\uparrow(z) \\ 0 \end{pmatrix}, \quad \Phi_3(z) = \begin{pmatrix} 0 \\ \varphi_\downarrow(z) \end{pmatrix},$$
$$\Phi_2(z) = \begin{pmatrix} \chi_\uparrow(z) \\ 0 \end{pmatrix}, \quad \Phi_4(z) = \begin{pmatrix} 0 \\ \chi_\downarrow(z) \end{pmatrix}, \tag{7.22}$$

It is straightforward to check the orthogonality of the functions (7.22), so with the proper choice of normalization constants in Eqs. (7.15), (7.19) we have

$$\int_{-L/2}^{L/2} \Phi_i^+(z)\Phi_j(z)dz = \delta_{ij}. \tag{7.23}$$

Orthogonality follows from the symmetry of the integrand. For example,

$$\Phi_1^+\Phi_2 = A_1^2 f_+(z,E_+)f_-(z,E_-) + B_1^2\eta_1(E_+)\eta_2(E_-)f_+(z,E_+)f_-(z,E_-), \tag{7.24}$$

It follows from (7.16) and (7.20) that the product (7.24) is the odd function of z and becomes zero after being integrated in the symmetric limits.

7.3 Effective Surface Hamiltonian

Orthonormal functions (7.22) can serve as a basis that can be used to calculate the effective Hamiltonian acting in a subspace of surface states. The average energy associated with surface states can be represented by the integral

$$\widetilde{H} = \int_{-L/2}^{L/2} \Psi^+(z)H\Psi(z)dz, \tag{7.25}$$

where H is given in Eq. (7.3) and the total wave function is the linear combination of the basis wave function (7.22):

$$\Psi(\mathbf{r}) = \frac{1}{\sqrt{V}} \sum_{\mathbf{k}} \sum_{i=1}^{4} a_{i\mathbf{k}} \Phi_i(z)\exp(i\mathbf{kr}), \tag{7.26}$$

where V is the volume of the film, and $\mathbf{k}$ is the two-dimensional vector parallel to the surface. Within the second quantization approach the wave function $\Psi(\mathbf{r})$ is considered to be an operator and coefficients $a_{i\mathbf{k}}(a^+_{i\mathbf{k}})$ are annihilation (creation) operators of the electrons in the four basis states and the momentum $\mathbf{k}$. Making use Eq. (7.26) in (7.25), we obtain the Hamiltonian

$$\widetilde{H} = \sum_{\mathbf{k}} \sum_{i,j=1}^{4} \widetilde{H}_{ij} a^+_{i\mathbf{k}} a_{j\mathbf{k}}, \quad \widetilde{H}_{ij} = \int_{-L/2}^{L/2} \Phi^+_i(z) H \Phi_j(z) dz. \tag{7.27}$$

The eigenvalues of matrix $\widetilde{H}_{ij}$ determine the surface energy spectrum in TI. The basis functions were chosen as eigenfunctions of the Hamiltonian H_0, so matrix $(\widetilde{H}_0)_{ij}$ is diagonal with surface energy levels on the main diagonal:

$$(\widetilde{H}_0)_{ij} = \mathrm{diag}\{E_+, E_-, E_+, E_-\} = E_0 + \mathrm{diag}\left\{\frac{\Delta_S}{2}, -\frac{\Delta_S}{2}, \frac{\Delta_S}{2}, -\frac{\Delta_S}{2}\right\},$$

$$E_0 = \frac{E_+ + E_-}{2}. \tag{7.28}$$

As the basis functions in Eq. (7.22) are written in the compact form using the quasi 2-spinors, then, in order to calculate $(\widetilde{H}_1)_{ij}$ in Eq. (7.27) it is convenient to present H_1 also in the compact 2 × 2 form as follows:

$$H_1 = [V_S(z) + V_{AS}(z)]\, I + \begin{pmatrix} -B_2 k^2 \tau_z & A_2 k_- \sigma_x \\ A_2 k_+ \sigma_x & -B_2 k^2 \tau_z \end{pmatrix}, \tag{7.29}$$

where τ_z and σ_x are the 2 × 2 Pauli matrices acting in two-band and two-spin space, respectively.

The calculation of the matrix element $(H_1)_{14}$ is given below as an example:

$$(\widetilde{H}_1)_{14} = \langle \Phi_1 | H_1 | \Phi_4 \rangle = \int_{-L/2}^{L/2} dz\, (\varphi^+_\uparrow(z), 0) \begin{pmatrix} -B_2 k^2 \tau_z & A_2 k_- \sigma_x \\ A_2 k_+ \sigma_x & -B_2 k_2 \tau_z \end{pmatrix} \begin{pmatrix} 0 \\ \chi_\downarrow(z) \end{pmatrix}$$

$$= A_2 k_- \langle \varphi_\uparrow | \sigma_x | \chi_\downarrow \rangle. \tag{7.30}$$

The matrix element $\langle \varphi_\uparrow | \sigma_x | \chi_\downarrow \rangle$ is imaginary

$$A_2\langle\varphi_\uparrow|\sigma_x|\chi_\downarrow\rangle = iA_1A_2\int_{-L/2}^{L/2}[\eta_1(E_+)f_-(z,E_-)f_- + \eta_2(E\_)f_+(z,E_-)f_+(z,E_+)]dz \equiv i\tilde{A}_2, \quad (7.31)$$

and it is finally expressed as

$$(\widetilde{H}_1)_{14} = i\widetilde{A}_2k_-. \quad (7.32)$$

Direct calculation of the matrix elements gives an effective Hamiltonian in the form

$$\widetilde{H} = \widetilde{H}_0 + \widetilde{H}_1 = (E_0 + \widetilde{V}_S + Dk^2)I + \begin{pmatrix} \frac{\Delta_S}{2} - Bk^2 & \widetilde{V}_{AS} & 0 & i\widetilde{A}_2k_- \\ \widetilde{V}_{AS} & -\frac{\Delta_S}{2} + Bk^2 & i\widetilde{A}_2k_- & 0 \\ 0 & -i\widetilde{A}_2k_+ & \frac{\Delta_S}{2} - Bk^2 & \widetilde{V}_{AS} \\ -i\widetilde{A}_2k_+ & 0 & \widetilde{V}_{AS} & -\frac{\Delta_S}{2} + Bk^2 \end{pmatrix}, \quad (7.33)$$

where

$$D = (\widetilde{B}_1 + \widetilde{B}_2)/2, B = (\widetilde{B}_1 - \widetilde{B}_2)/2,$$
$$\widetilde{B}_1 = B_2\langle\varphi_\uparrow|\tau_z|\varphi_\uparrow\rangle, \widetilde{B}_2 = B_2\langle\chi_\uparrow|\tau_z|\chi_\uparrow\rangle,$$
$$\widetilde{V}_{AS} = \langle\varphi_\uparrow|V_{AS}(z)|\chi_\uparrow\rangle = \langle\varphi_\downarrow|V_{AS}(z)|\chi_\downarrow\rangle,$$
$$\widetilde{V}_S = \langle\varphi_\uparrow|V_S(z)|\varphi_\uparrow\rangle = \langle\chi_\uparrow|V_S(z)|\chi_\uparrow\rangle. \quad (7.34)$$

Finally, we obtain four branches of the surface energy spectrum as the solution of the equation $det(\widetilde{H}_{ij} - E) = 0$:

$$E_{c,v\uparrow}(k) = E_0 + \widetilde{V}_S + Dk^2 \pm R_\uparrow, E_{c,v\downarrow}(k) = E_0 + \widetilde{V}_S + Dk^2 \pm R_\downarrow,$$
$$R_{\uparrow\downarrow} = \sqrt{\left(\frac{\Delta_S}{2} - Bk^2\right)^2 + (\widetilde{A}_2k \mp \widetilde{V}_{AS})^2}, \quad k^2 = k_x^2 + k_y^2. \quad (7.35)$$

Signs ± in front of the square roots describe conduction and valence bands, and signs under the roots correspond to the spin variable, ↑↓. Spin states ↑↓ are the mixtures of pure spinors in the bulk. Spectrum (7.35) describes real surface electrons that can be experimentally probed with the angle-resolved photoemission spectroscopy.

Surface bands (7.35) are spin-split if the spin-orbit interaction ($\widetilde{A}_2 \neq 0$) and the structural inversion asymmetry ($V_{AS} \neq 0$) act simultaneously. This lifts the spin degeneracy so the bands are Kramers degenerate $E_{\uparrow}(k) = E_{\downarrow}(-k)$ as required by the time reversal symmetry. The surface electrons behave like the two-dimensional Rashba electron gas discussed in Chapter 2. The spectrum is sensitive to the relative sign of the gap and the effective mass. If the surface spectrum has a direct gap ($B\Delta_s < 0$), the energy dispersion has the shape shown in Fig. 7.6.

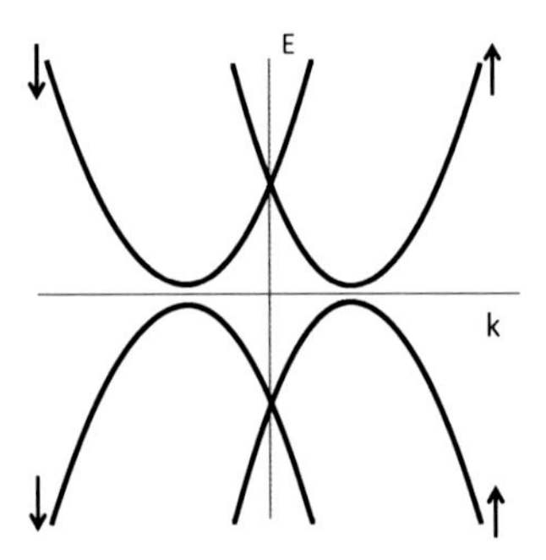

Figure 7.6 Surface energy spectrum, $B\Delta_s < 0$. Arrows indicate spin variables.

If surface bands are inverted, $B\Delta_s > 0$, the dispersion changes as illustrated in Fig. 7.7.

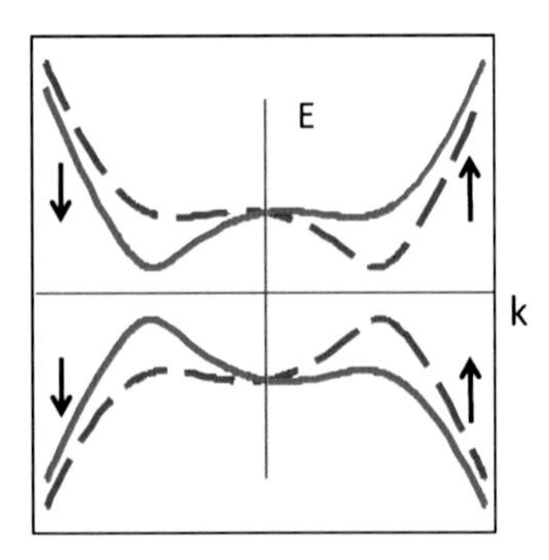

Figure 7.7 Inverted surface energy spectrum, $B\Delta_s < 0$. Dashed and solid lines illustrate branches with opposite spins.

The energy bands in Figs. 7.6 and 7.7 are shown with no indication of the surface they belong to. Electron localization near the top and bottom surfaces will be discussed below.

Eigenfunctions of the Hamiltonian (7.33) normalized to unity are given below:

$$C_{\uparrow\downarrow} = \sqrt{\frac{Bk^2 - \Delta_S/2 + R_{\uparrow\downarrow}}{4\,R_{\uparrow\downarrow}}} \begin{pmatrix} \dfrac{(\widetilde{A}_2 k \mp \widetilde{V}_{AS})\, i\exp(-i\varphi)}{Bk^2 - \Delta_S/2 + R_{\uparrow\downarrow}} \\ \mp i\exp(-i\varphi) \\ \dfrac{\widetilde{V}_{AS} \mp \widetilde{A}_2 k}{Bk^2 - \Delta_S/2 + R_{\uparrow\downarrow}} \\ 1 \end{pmatrix},$$

$$V_{\uparrow\downarrow} = \sqrt{\frac{\Delta_S/2 - Bk^2 + R_{\uparrow\downarrow}}{4\,R_{\uparrow\downarrow}}} \begin{pmatrix} \dfrac{-(\widetilde{A}_2 k \mp \widetilde{V}_{AS})\, i\exp(-i\varphi)}{\Delta_S/2 - Bk^2 + R(\uparrow\downarrow)} \\ \mp i\exp(-i\varphi) \\ -\dfrac{\widetilde{V}_{AS} \mp \widetilde{A}_2 k}{\Delta_S/2 - Bk^2 + R_{\uparrow\downarrow}} \\ 1 \end{pmatrix},$$

$$\tan(\phi) = \frac{k_y}{k_x}. \tag{7.36}$$

The upper (lower) sign in Eq. (7.36) corresponds to spin, $S = \uparrow(\downarrow)$. It is straightforward to check that the spinors are orthonormal

$$C_S^+ C_{S'} = V_S^+ V_{S'} = \delta_{SS'}, \quad C_S^+ V_{S'} = 0, \tag{7.37}$$

and the matrix Hamiltonian calculated on the new basis (7.36) becomes diagonal. Energy bands (7.35) can be found as the matrix elements

$$E_{c,S}(k) = \langle C_S^+ | \tilde{H} | C_S \rangle$$
$$E_{v,S}(k) = \langle V_S^+ | \tilde{H} | V_S \rangle. \tag{7.38}$$

It is convenient to use the basis that diagonalizes the Hamiltonian (7.33) to study the effects of an external field applied to experimentally observable surface electrons. As an example, in Chapter 8 we discuss how surface electrons interact with magnetic impurities.

7.4 Spatial Distribution of Surface Electrons

Let us consider the spatial dependence of the surface states when $k = 0$, $\tilde{V}_{AS} = 0$. The Hamiltonian (7.33) is diagonal and Eq. (7.22) represents its eigenfunctions. The electron probability density function $\Phi_1^{+}(z)\Phi_1(z)$ normalized to unity is shown in Fig. 7.8. Calculation was performed with the set of parameters given in Ref. [7]: $A_1 = 3.3$ eVÅ, $B_1 = 1.5$ eVÅ^2, $\Delta = 0.28$ eV.

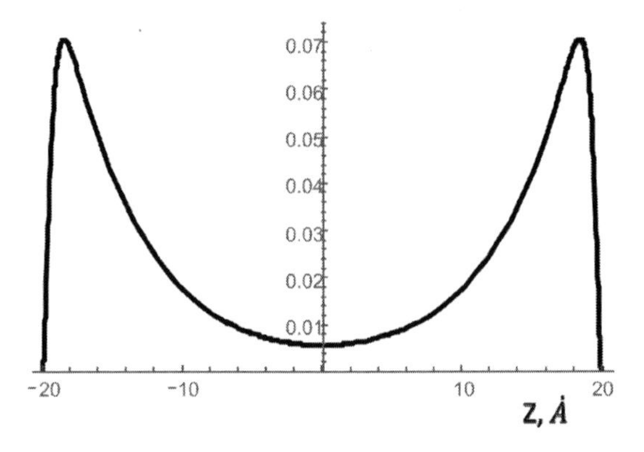

Figure 7.8 Electron probability density $\Phi_1^{+}(z)\Phi_1(z)$ across a 40 Å-thick TI film.

Since wave functions $\Phi_1(z)$ and $\Phi_2(z)$ correspond to surface energy levels with a gap between them, we refer to them as surface valence and conduction states, respectively. Functions that localize near the top $F_{\mathrm{T}}(z)$ and bottom $F_{\mathrm{B}}(z)$ surfaces can be represented as linear combinations of conduction and valence states:

$$F_{T\uparrow}(z) = \frac{1}{\sqrt{2}}\left(\Phi_1(z) - \Phi_2(z)\right)$$

$$F_{B\uparrow}(z) = \frac{1}{\sqrt{2}}\left(\Phi_1(z) + \Phi_2(z)\right), \tag{7.39}$$

Probability densities are shown in Fig. 7.9.

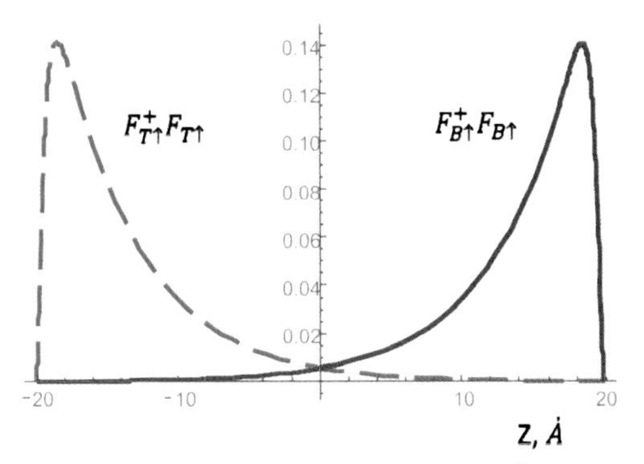

Figure 7.9 Top and bottom electrons in a 40 Å thick film. Positive z-direction points down to bottom surface.

Top and bottom functions $F_{T\uparrow}(z)$, $F_{B\uparrow}(z)$ can serve as a basis only approximately, to the accuracy that neglects their overlap in the center of the film. In other words, the top and bottom are good quantum numbers if the film thickness is large enough.

Using the eigenstates (7.36) one can compose the z-dependent linear combination of basis wave functions $\Phi_{1-4}(z)$ which makes the Hamiltonian (7.33) diagonal with eigenvalues (7.35) on its main diagonal:

$$
\begin{aligned}
V_\uparrow(z) &= -\frac{(\tilde{A}_2k - \tilde{V}_{AS})\, i\exp(-i\varphi)}{\Delta_S/2 - Bk^2 + R_\uparrow}\Phi_1(z) - i\exp(-i\varphi)\,\Phi_2(z) \\
&\quad + \frac{\tilde{A}_2k - \tilde{V}_{AS}}{\Delta_S/2 - Bk^2 + R_\uparrow}\Phi_3(z) + \Phi_4(z), \\
C_\uparrow(z) &= \frac{(\tilde{A}_2k - \tilde{V}_{AS})\, i\exp(-i\varphi)}{Bk^2 - \Delta_S/2 + R_\uparrow}\Phi_1(z) - i\exp(-i\varphi)\Phi_2(z) \\
&\quad + \frac{\tilde{V}_{AS} - \tilde{A}_2k}{Bk^2 - \Delta_S/2 + R_\uparrow}\Phi_3(z) + \Phi_4(z), \\
V_\downarrow(z) &= -\frac{(\tilde{A}_2k + \tilde{V}_{AS})\, i\exp(-i\varphi)}{\Delta_S/2 - Bk^2 + R_\downarrow}\Phi_1(z) + i\exp(-i\varphi)\Phi_2(z) \\
&\quad - \frac{\tilde{V}_{AS} + \tilde{A}_2k}{\Delta_S/2 - Bk^2 + R_\downarrow}\Phi_3(z) + \Phi_4(z), \\
C_\downarrow(z) &= \frac{(\tilde{A}_2k + \tilde{V}_{AS})\, i\exp(-i\varphi)}{Bk^2 - \Delta_S/2 + R_\downarrow}\Phi_1(z) + i\exp(-i\varphi)\Phi_2(z) \\
&\quad + \frac{\tilde{V}_{AS} + \tilde{A}_2k}{Bk^2 - \Delta_S/2 + R_\downarrow}\Phi_3(z) + \Phi_4(z),
\end{aligned}
\tag{7.40}
$$

Wave functions (7.40) reside at the top or bottom surfaces, depending on the in-plane wave vector ($\pm k$) and the sign of the factor $\tilde{A}_2 k + \tilde{V}_{AS}$. An example is shown in Figs. 7.10 and 7.11, where the valence (conduction) electron density at small k ($\tilde{A}_2 k < \tilde{V}_{AS}$) is located predominantly at the top (bottom) surface.

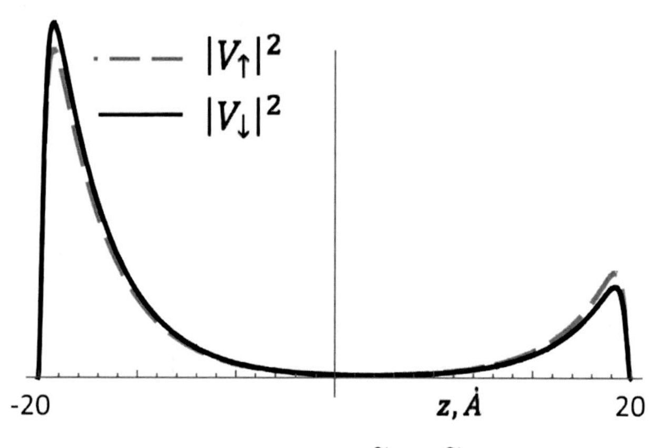

Figure 7.10 Valence probability density, $\tilde{A}_2 k < \tilde{V}_{AS}$.

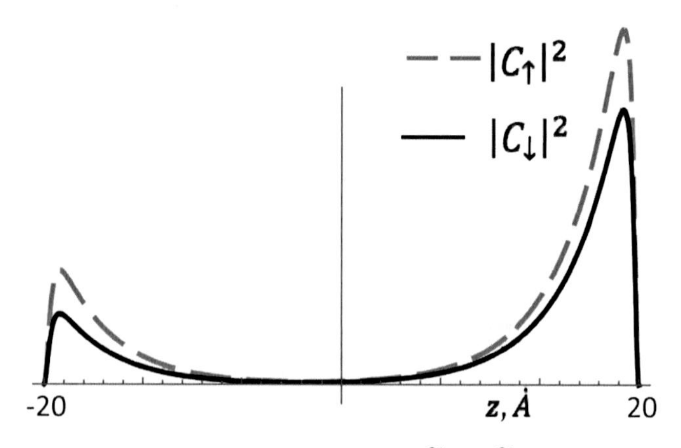

Figure 7.11 Conduction probability density, $\tilde{A}_2 k < \tilde{V}_{AS}$.

The same analyses can be performed for large k, ($\tilde{A}_2 k > \tilde{V}_{AS}$) and also for the states at $-k$. Results are shown in Fig. 7.12 where the electron energy spectrum from Fig. 7.6 is detailed as to indicate the electron localization.

If an external field changes its sign the top and bottom branches in Fig. 7.12 swap their places.

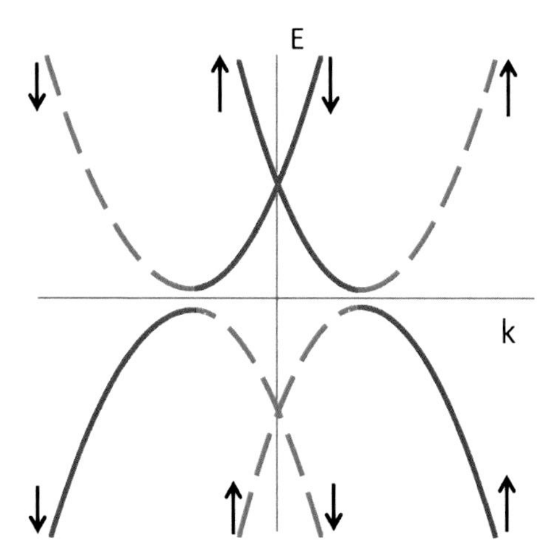

Figure 7.12 Electron spectrum (7.35). Solid (bottom) and dashed (top) lines indicate electron localization near opposite surfaces.

At small k the spectrum (7.35) is k-linear

$$E_{c,v\uparrow\downarrow} = E_0 + \widetilde{V}_S \pm \left[\widetilde{\Delta} \pm \frac{\widetilde{A}_2 \widetilde{V}_{AS} k}{\widetilde{\Delta}}\right],$$

$$\widetilde{\Delta} = \left(\frac{\Delta_S^2}{4} + \widetilde{V}_{AS}^2\right)^{1/2}. \tag{7.41}$$

Signs ± in front of the square brackets in Eq. (7.41) correspond to conduction and valence electrons, respectively. Dirac cones from (7.41) are shown in Fig. 7.13.

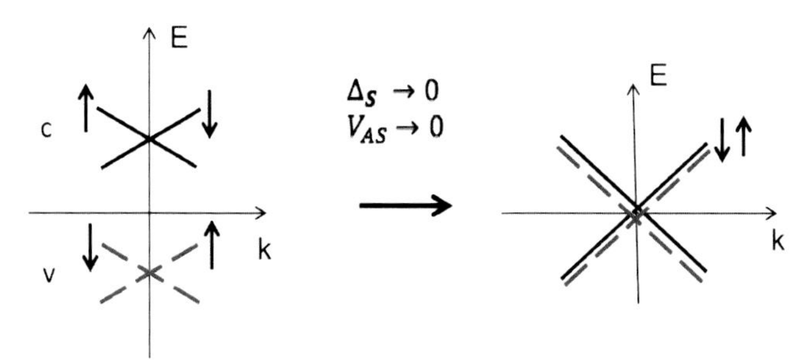

Figure 7.13 Spin polarized Dirac cones. Dashed and solid lines differentiate the energy dispersions of electrons located near opposite surfaces.

The limit shown in Fig. 7.13 means that the thickness is large enough to neglect tunneling across the film, $\Delta_S \to 0$, and also the asymmetric external field tends to zero, $\widetilde{V}_{AS} \to 0$. In this limit the energy branches become spin degenerate.

Once we know spinor eigenstates (7.36) and corresponding energy branches (7.35), we may use them to study the indirect exchange interaction between magnetic impurities placed in TI. The indirect exchange mediated by surface electrons is calculated in Chapter 8.

7.5 Topological Invariant

The topological properties of TI are defined by their symmetry with respect to the time reversal operation. The time reversal operator acting on a single particle with spin $s = 1/2$ can be represented by the matrix

$$\Theta = -i\sigma_y K = \begin{pmatrix} 0 & -1 \\ 1 & 0 \end{pmatrix} K, \tag{7.42}$$

where σ_y is the Pauli matrix and K is the operator of complex conjugation. Acting on pure spin states $\begin{pmatrix} 1 \\ 0 \end{pmatrix}$ and $\begin{pmatrix} 0 \\ 1 \end{pmatrix}$ the operator Θ flips the spin direction:

$$\Theta\begin{pmatrix} 1 \\ 0 \end{pmatrix} = \begin{pmatrix} 0 \\ 1 \end{pmatrix}, \ \Theta\begin{pmatrix} 0 \\ 1 \end{pmatrix} = -\begin{pmatrix} 1 \\ 0 \end{pmatrix}, \tag{7.43}$$
$$\Theta^2 = -1.$$

In our four-band model, the sequence of bands we choose in the Hamiltonian (7.33) dictates the form of the time reversal operator that follows:

$$\Theta = \begin{pmatrix} 0 & 0 & -1 & 0 \\ 0 & 0 & 0 & -1 \\ 1 & 0 & 0 & 0 \\ 0 & 1 & 0 & 0 \end{pmatrix} K. \tag{7.44}$$

Let us inspect the behavior of wave functions (7.36) under the operation Θ:

$$\Theta C_\uparrow(k) = i\exp(i\varphi)\sqrt{\frac{Bk^2 - \Delta_S/2 + R_\uparrow}{4R_\uparrow}}\begin{pmatrix} \dfrac{(-\widetilde{A}_2k + \widetilde{V}_{AS})\, i\exp(-i\varphi)}{Bk^2 - \Delta_S/2 + R_\uparrow} \\ i\exp(-i\varphi) \\ \dfrac{\widetilde{V}_{AS} - \widetilde{A}_2k}{Bk^2 - \Delta_S/2 + R_\uparrow} \\ 1 \end{pmatrix}$$

$$= i\exp(i\varphi)\, C_\downarrow(-k)$$

$$\Theta C_\downarrow(k) = -i\exp(i\varphi)\sqrt{\frac{Bk^2 - \Delta_S/2 + R_\downarrow}{4R_\downarrow}}\begin{pmatrix} -\dfrac{(\widetilde{A}_2k + \widetilde{V}_{AS})\; i\exp(-i\varphi)}{Bk^2 - \Delta_S/2 + R_\downarrow} \\ -i\exp(-i\varphi) \\ \dfrac{\widetilde{A}_2k + \widetilde{V}_{AS}}{Bk^2 - \Delta_S/2 + R_\downarrow} \\ 1 \end{pmatrix}$$

$$= -i\exp(i\varphi) C_\uparrow(-k)$$

$$\Theta V_\uparrow(k) = i\exp(i\varphi)\sqrt{\frac{\Delta_S/2 - Bk^2 + R_\uparrow}{4R_\uparrow}}\begin{pmatrix} -\dfrac{(\widetilde{A}_2k - \widetilde{V}_{AS})i\exp(-i\varphi)}{\Delta_S/2 - Bk^2 + R_\uparrow} \\ i\exp(-i\varphi) \\ \dfrac{\widetilde{A}_2k - \widetilde{V}_{AS}}{\Delta_S/2 - Bk^2 + R_\uparrow} \\ 1 \end{pmatrix}$$

$$= i\exp(i\varphi)\, V_\downarrow(-k)$$

$$\Theta V_\downarrow(k) = -i\exp(i\varphi)\sqrt{\frac{\Delta_S/2 - Bk^2 + R_\downarrow}{4R_\downarrow}}\begin{pmatrix} \dfrac{(\widetilde{A}_2k + \widetilde{V}_{AS})i\exp(-i\varphi)}{\Delta_S/2 - Bk^2 + R_\downarrow} \\ -i\exp(-i\varphi) \\ -\dfrac{\widetilde{A}_2k + \widetilde{V}_{AS}}{\Delta_S/2 - Bk^2 + R_\downarrow} \\ 1 \end{pmatrix}$$

$$= -i\exp(i\varphi)\, V_\uparrow(-k) \tag{7.45}$$

Using (7.45) it is straightforward to obtain the matrix overlap of Kramers conjugate partners

$$\langle i|\Theta|j\rangle = \begin{pmatrix} 0 & 0 & a_{13} & a_{14} \\ 0 & 0 & a_{23} & a_{24} \\ -a_{13} & -a_{23} & 0 & 0 \\ -a_{14} & -a_{24} & 0 & 0 \end{pmatrix}, i, j = (1, 2, 3, 4) \to (V\uparrow, C\uparrow, V\downarrow, C\downarrow)$$

$$a_{13} = i\exp(i\varphi)\frac{\tilde{A}_2^2k^2 - \tilde{V}_{AS}^2 - (\Delta_S/2 - Bk^2 + R_\uparrow)(\Delta_S/2 - Bk^2 + R_\downarrow)}{2\sqrt{R_\uparrow R_\downarrow}\sqrt{\Delta_S/2 - Bk^2 + R_\uparrow}\sqrt{\Delta_S/2 - Bk^2 + R_\downarrow}},$$

$$a_{24} = i\exp(i\varphi)\frac{\tilde{A}_2^2k^2 - \tilde{V}_{AS}^2 - (Bk^2 - \Delta_S/2 + R_\uparrow)(Bk^2 - \Delta_S/2 + R_\downarrow)}{2\sqrt{R_\uparrow R_\downarrow}\sqrt{Bk^2 - \Delta_S/2 + R_\uparrow}\sqrt{Bk^2 - \Delta_S/2 + R_\downarrow}},$$

$$a_{14} = i\exp(i\varphi)\frac{\tilde{V}_{AS}^2 - \tilde{A}_2^2k^2 - (\Delta_S/2 - Bk^2 + R_\uparrow)(Bk^2 - \Delta_S/2 + R_\downarrow)}{2\sqrt{R_\uparrow R_\downarrow}\sqrt{\Delta_S/2 - Bk^2 + R_\uparrow}\sqrt{Bk^2 - \Delta_S/2 + R_\downarrow}},$$

$$a_{23} = i\exp(i\varphi)\frac{\tilde{V}_{AS}^2 - \tilde{A}_2^2k^2 - (Bk^2 - \Delta_S/2 + R_\uparrow)(\Delta_S/2 - Bk^2 + R_\downarrow)}{2\sqrt{R_\uparrow R_\downarrow}\sqrt{Bk^2 - \Delta_S/2 + R_\uparrow}\sqrt{\Delta_S/2 - Bk^2 + R_\downarrow}}. \quad (7.46)$$

We assume that the chemical potential is in the surface energy gap and we express the overlap of filled (valence) bands as a submatrix of (7.46):

$$\langle v|\Theta|v\rangle = \begin{pmatrix} 0 & \alpha_{13} \\ -\alpha_{13} & 0 \end{pmatrix}, \quad (7.47)$$

where $P(k) = \sqrt{\mathrm{Det}\langle[v|\Theta|v]\rangle} = a_{13}$ is the Pfaffian to the matrix (7.47). As was shown in Ref. [2] the topological invariant that distinguishes the non-trivial state of TI from an ordinary dielectric is the number of pairs of zeroes in $P(k)$ along the contour passing the Brillouin zone as shown in Fig. 7.14 for a simple square lattice.

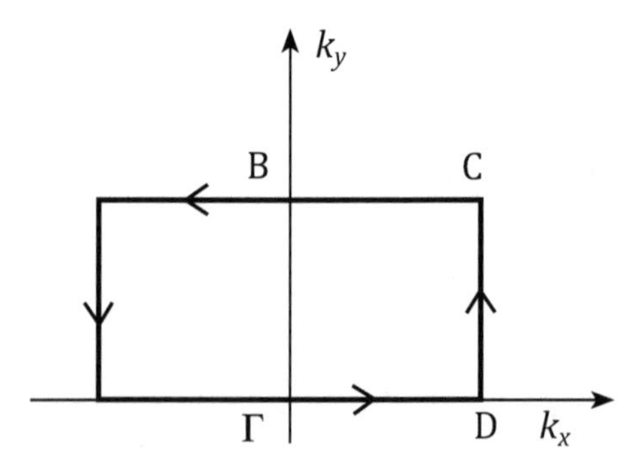

Figure 7.14 Path to follow the winding of the phase of $P(k)$. A loop encloses half of the Brillouin zone.

Points A, B, C, D in Fig. 7.14 are the time reversal invariant momenta (TRIM), where $|P(k)| = 1$. Points $B(0, G/2)$, $C(G/2, G/2)$, and $D(G/2, 0)$ lie on the edge of the first Brilluoin zone, $G = 2\pi/a$, a is the lattice constant. The number of zeroes I is half the number of $P(k)$ sign changes along the path, and I modulo 2 is called the Z_2 topological index, so it equals 1 or 0 when I is odd or even, respectively. One pair of zeroes means that $P(k)$ changes its sign twice, so Z_2 index $I = 1$ characterizes a topologically non-trivial state. The state with $I = 0$ corresponds to that of a simple dielectric.

In order to illustrate the rule let us assume $\widetilde{V}_{AS} = 0$ and analyze $P(k)$ as given in Eq. (7.46). Since we are dealing with a continuous model, all three points B, C, and D move to infinity in the limit of lattice constant $a \to 0$, and the change of the sign of $P(k)$ can be followed on a k-half-axis $(0, \infty)$. It is easy to check that $P(0) = -\text{sign}\,(\Delta_S)$, $P(\infty) = \text{sign}(B)$. The graph of $P(k)$ is shown in Fig. 7.15 where numerical data are taken from Ref. [7]: $B = 10$ eVÅ^2, $\Delta_S = 0.07$ eV, $A_2 \approx 4$ eVÅ.

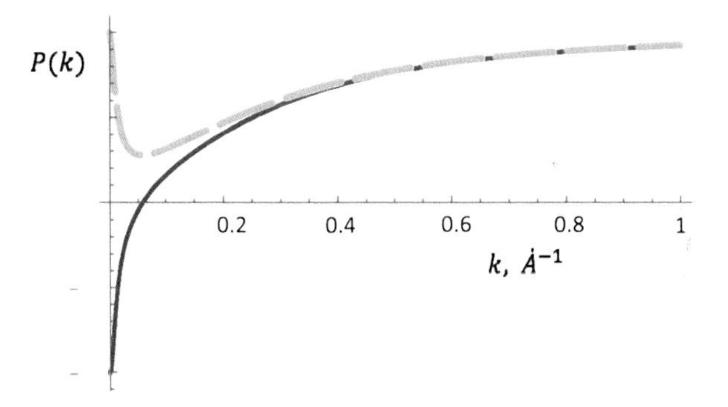

Figure 7.15 $P(k)$ as a function of k. Solid line: $B\Delta_S > 0$, dashed line: $B\Delta_S < 0$.

If $B\Delta_S > 0$, there is one sign change in $P(k)$ as illustrated in Fig. 7.15 manifesting the non-trivial or quantum spin Hall phase. If $B\Delta_S < 0$ the sign change is absent and thus the direct gap surface bands are in a trivial semiconductor state.

Problems

7.1 Find the optical selection rules for the transitions between surface states at $k = 0$, (7.22).

7.2 Find optical matrix elements for top-bottom transitions.

References

1. Zhu ZH, Levy G, Ludbrook B, Veenstra CN, Rosen JA, Comin R, Wong D, Dosanjh P, Ubaldini A, Syers P, Butch NP, Paglione J, Elfimov IS, A. Damascelli A (2011) Rashba spin-splitting control at the surface of the topological insulator Bi_2Se_3, *Phys Rev Lett*, **107**, 186405.
2. Kane CL, Mele EJ (2005) Z_2 topological order and the quantum spin Hall effect, *Phys Rev Lett*, **95**, 146802.
3. Bernevig BA, Huges TL, Zhang SC (2006) Quantum spin Hall effect and topological phase transition in HgTe quantum wells, *Science*, **314**, 1757–1761; Konig M, Wiedmann S, Brüne C, Roth A, Buhmann H, Molenkamp LW, Qi XL, Zhang SC (2007) Quantum spin Hall insulator state in HgTe quantum wells, *Science*, **318**, 766–770.
4. Hasan MZ, Kane CL (2010) Colloquium: Topological insulators, *Rev Mod Phys*, **82**, 3045–3067; Qi XL, Zhang SC (2011) Topological insulators and superconductors, *Rev Mod Phys*, **83**, 1057–1110.
5. Brune C, Roth A, Buhmann H, Hankiewicz EM, Molenkamp LW, Maciejko J, Qi X-L, Zhang S-C (2012) Spin polarization of the spin Hall edge states, *Nat Phys*, **8**, 485–490.
6. Haijun Zhang H, Liu CX, Qi XL, Dai X, Fang Z, Zhang SC (2009) Topological insulators in Bi_2Se_3, Bi_2Te_3 and Sb_2Te_3 with a single Dirac cone on the surface, *Nat Phys*, **5**, 438–442.
7. Shan W-Y, Lu H-Z, Shen S-Q (2010) Effective continuous model for surface states and thin films of three-dimensional topological insulators, *N J Phys*, **12**, 043048; Lu H-Z, Shan W-Y, Wang Y, Niu Q, Shen S-Q (2010) Massive Dirac fermions and spin physics in an ultrathin film of topological insulator, *Phys Rev B*, **81**, 115407.
8. Miao MS, Yan Q, Van de Walle CG, Lou WK, Li LL, Chang K (2012) Polarization-driven topological insulator transition in a GaNInNGaN quantumwell, *Phys Rev Lett*, **109**, 186803.
9. Liang Fu L (2011) Topological crystalline insulators, *Phys Rev Lett*, **106**, 106802.
10. Tanaka Y, Ren Z, Sato T, Nakayama K, Souma S, Takahashi T, Segawa K, Ando Y (2012) Experimental realization of a topological crystalline insulator in SnTe, *Nat Phys*, **8**, 800–803.
11. Hsieh TH, Lin H, Liu J, Duan W, Bansil A, Fu L (2012) Topological crystalline insulators in the SnTe material class, *Nat Commun*, **3**, Article number: 982.

Chapter 8

Magnetic Exchange Interaction in Topological Insulator

The subject of the topological protection of surface states against momentum scattering events and strong coupling between electron spin and momentum has attracted considerable attention to possible spintronic applications of topological insulators (TI). External magnetic fields break the time reversal symmetry that turns a topological state on and off [1]. An alternative way to bring TI to a trivial state is magnetic doping [2] that forms a ferromagnetic order on the surface. Developing new ways to electrically control surface states would be a critical step toward the practical exploitation of their topological properties.

The way surface ferromagnetism acts on TI is that it opens an energy gap in the Dirac spectrum on a single surface. Ferromagnetism itself is affected by the surface electron spectrum and the symmetry of surface wave functions as it may or may not develop depending on indirect exchange interaction mediated by surface electrons and holes. One of the approaches to the problem is based on the model that describes massless Dirac electrons interacting with localized spins by a contact *s*–*d* interaction [3–8]. This model is simplistic as it comprises two terms of different origins: the Dirac model which is an effective Hamiltonian

Wide Bandgap Semiconductor Spintronics
Vladimir Litvinov

ISBN 978-981-4669-70-2 (Hardcover), 978-981-4669-71-9 (eBook)
www.panstanford.com

obtained by a projection of the bulk one onto surface states (see Chapter 7), and the magnetic *s–d* interaction which is postulated to be equivalent to the one in the bulk. As a result, the *s–d* interaction constants do not contain any information on surface states and then do not depend on the position of localized spin relative to the surface. Also, as formulated, the model describes a single Dirac cone and cannot be applied to a film where the electron states on opposite surfaces couple to each other.

A consistent approach to surface magnetism should operate with the effective *s–d* coupling calculated as a projection of the bulk Hamiltonian that comprises electrons, localized spins, and the contact *s–d* interaction between them. In this chapter, we construct an effective surface model which serves as a background for the analytical study of surface magnetism. It is shown that in this effective surface model, the interaction matrix depends on the position of localized spins relative to the slab surface. It results in different impurity spin textures if we move away from the surface. Also, *s–d* interaction constants depend on the parameters of an electron spectrum that make the constants specific to a particular TI whether it is a thick sample with a single massless Dirac cone or a thin slab with gapped massive fermions. Third, the interaction matrix is a function of the gate bias applied across the slab, and is affected by the Rashba spin splitting experimentally observed in TI [9]. Finally, this effective surface model predicts TI properties which cannot follow from the simplified model, especially when external bias is applied [10]. It is shown that an effective *s–d* exchange is sensitive to the symmetry of surface wave functions and depends on the position of an impurity spin relative to slab surfaces. Under applied voltage the indirect exchange oscillating range function reveals zero magnetic field beating as a signature of Rashba spin splitting.

8.1 Spin-Electron Interaction

An electron interacts with a magnetic impurity spin located at point $\mathbf{R}_0$ through the contact *s–d* interaction

$$H_{sd} = \frac{J}{n}\mathbf{S}\sigma\,\delta(\mathbf{r} - \mathbf{R}_0), \tag{8.1}$$

where n is the volume density of host atoms, **S** and $\boldsymbol{\sigma}$ are the spin operators of impurity and electron, respectively. In a single-band metal, the s–d interaction can be represented in a second quantization form by expanding the operator wave function in plane waves normalized on a crystal volume

$$\Psi(\mathbf{r}) = \frac{1}{\sqrt{V}} \sum_{\mathbf{k},s} a_{\mathbf{k}s} \exp(i\mathbf{kr}), \tag{8.2}$$

and then calculating the matrix element of the interaction (8.1):

$$\widetilde{H}_{sd} = \int \Psi^{+}(\mathbf{r}) H_{sd} \Psi(\mathbf{r})\, d\mathbf{r} = \frac{J}{N} \sum_{\mathbf{k},\mathbf{q}} \sum_{s,s'} \exp(i\mathbf{qR}_0) \mathbf{S}\boldsymbol{\sigma}_{s,s'} a^{+}_{\mathbf{k}s} a_{\mathbf{k}+q,s'} \tag{8.3}$$

where $a_{\mathbf{k}s}$ is the annihilation operator of an electron with spin $s = \pm 1/2$, $\boldsymbol{\sigma}_{s,s'}$ are Pauli matrices, N is the number of host atoms.

Interaction (8.3) regarded as a perturbation up to the second order, generates the indirect exchange coupling between a pair of impurity spins. In a simple metal that is the RKKY interaction (see Chapter 5). In semiconductor host Bi_2Te_3 the basis wave functions are more complex than the single-band plane waves used in (8.3). As discussed in Chapter 7, the set of basis wave functions in bulk Bi_2Te_3 is the 4-spinor in band and spin indexes. The basis of the representation is given below:

$$\Psi(\mathbf{r}) = \frac{1}{\sqrt{A}} \sum_{\mathbf{k}j} a_{\mathbf{k}j} u_j \exp(i\mathbf{kr}_{\parallel}),\ \mathbf{r}_{\parallel} = (x, y),\ \mathbf{k} = (k_x, k_y), \tag{8.4}$$

where u_j are the Bloch amplitudes at $\mathbf{k} = 0$.

The free electron Hamiltonian in an external electric field has the form

$$H_F = \sum_{i,j} H_{ij}(\mathbf{k})\, a^{+}_{i\mathbf{k}} a_{j\mathbf{k}}, \tag{8.5}$$

and quantum numbers i, j run over conduction/valence bands and spins in a sequence, $i, j = 1(v\uparrow), 2(c\uparrow), 3(v\downarrow), 4(c\downarrow)$:

$$H(\boldsymbol{k}) = H_0 + H_1,$$

$$H_0 = \begin{pmatrix} -\Delta + B_1 \dfrac{\partial^2}{\partial z^2} & -iA_1 \dfrac{\partial}{\partial z} & 0 & 0 \\ -iA_1 \dfrac{\partial}{\partial z} & \Delta - B_1 \dfrac{\partial^2}{\partial z^2} & 0 & 0 \\ 0 & 0 & -\Delta + B_1 \dfrac{\partial^2}{\partial z^2} & iA_1 \dfrac{\partial}{\partial z} \\ 0 & 0 & iA_1 \dfrac{\partial}{\partial z} & \Delta - B_1 \dfrac{\partial^2}{\partial z^2} \end{pmatrix},$$

$$H_1 = [V_S(z) + V_{AS}(z)]\, I + \begin{pmatrix} -B_2k^2 & 0 & 0 & A_2k_- \\ 0 & B_2k^2 & A_2k_- & 0 \\ 0 & A_2k_+ & -B_2k^2 & 0 \\ A_2k_+ & 0 & 0 & B_2k^2 \end{pmatrix}. \tag{8.6}$$

In order to compose the full Hamiltonian including magnetic interaction we have to express the *s*-*d* coupling (8.1) in the same representation as the Hamiltonian (8.6):

$$H_{sd} = \int \Psi^+(\mathbf{r}_{||})\, H_{sd} \Psi(\mathbf{r}_{||}) d\mathbf{r}_{||}$$

$$= \frac{1}{nA} \sum_{\mathbf{k}i,\mathbf{k}'j} J_{ij} \exp(i(\mathbf{k}-\mathbf{k}')R_{0||}))\delta(z-Z) a^+_{i\mathbf{k}} a_{j\mathbf{k}'},$$

$$J_{ij} = \begin{pmatrix} J_v S_z & 0 & J_{||v} S_- & 0 \\ 0 & J_c S_z & 0 & J_{||c} S_- \\ J_{||v} S_+ & 0 & -J_v S_z & 0 \\ 0 & J_{||c} S_+ & 0 & -J_c S_z \end{pmatrix}, \tag{8.7}$$

where A is the film area. The z-coordinate is kept separate to account for finite size effects in z-direction, $S_\pm = S_x \pm iS_y$, $Z \equiv R_{0z}$. The interaction matrix (8.7) is obtained under the assumption that interband matrix elements are much smaller than intraband ones and can be neglected. One more simplification is possible if energy bands possess electron-hole symmetry. This reduces the number of coupling parameters: $J_v = J_c \equiv J_z, J_{||v} = J_{||c} \equiv J_{||}$.

It is shown in Chapter 7 how using surface basis wave functions (7.23) and matrix elements (7.26) one could get an effective Hamiltonian for surface states. The same procedure applied to the full Hamiltonian $H_F + H_{sd}$ gives the full effective Hamiltonian that includes surface electrons and their interaction with an impurity spin:

$$H_S = \int_{-L/2}^{L/2} \Phi_i^+ (H_F + H_{sd})\Phi_j dz = H_S^0 + H_S^1 + H_S^{sd} \tag{8.8}$$

The first two terms in Eq. (8.8) were obtained in Chapter 7:

$$H_S^0 = E_0 + Dk^2 + V_S,$$

$$H_S^1 = \begin{pmatrix} \Delta_S/2 - Bk^2 & V_{as} & 0 & i\widetilde{A}_2 k_- \\ V_{as} & -\Delta_S/2 + Bk^2 & i\widetilde{A}_2 k_- & 0 \\ 0 & -i\widetilde{A}_2 k_+ & \Delta_S/2 - Bk^2 & V_{as} \\ -i\widetilde{A}_2 k_+ & 0 & V_{as} & -\Delta_S/2 + Bk^2 \end{pmatrix}. \tag{8.9}$$

In order to find the surface *s*-*d* interaction

$$H_S^{sd} = \int_{-L/2}^{L/2} \Phi_i^+ H_{sd} \Phi_j dz, \tag{8.10}$$

it is convenient to represent (8.7) in 2 × 2 spin matrix form, where each matrix element is the 2 × 2 matrix in band indexes:

$$J_{ij} = \begin{pmatrix} X_{11} & X_{21} \\ X_{12} & X_{22} \end{pmatrix},$$

$$X_{11} = \begin{pmatrix} J_v S_z & 0 \\ 0 & J_c S_z \end{pmatrix}, X_{22} = \begin{pmatrix} -J_v S_z & 0 \\ 0 & -J_c S_z \end{pmatrix},$$

$$X_{12} = \begin{pmatrix} J_{||v} S_- & 0 \\ 0 & J_{||c} S_- \end{pmatrix}, X_{21} = \begin{pmatrix} J_{||v} S_+ & 0 \\ 0 & J_{||c} S_+ \end{pmatrix}. \tag{8.11}$$

After that the effective Hamiltonian (8.10) can be calculated directly making use of wave functions given in Chapter 7, Eq. (7.22):

$$H_S^{sd} = \frac{1}{nA} \sum_{i\mathbf{k}, j\mathbf{k}'} W_{ij}(Z) \exp(i(\mathbf{k} - \mathbf{k}')\mathbf{R}_{0||}) a_{i\mathbf{k}}^+ a_{j\mathbf{k}'}, \tag{8.12}$$

where

$$W(Z)=\begin{pmatrix} \tilde{J}_z S_z & \tilde{J}_{12} S_z & \tilde{J}_{||} S_- & 0 \\ \tilde{J}_{12} S_z & \tilde{J}_z S_z & 0 & -\tilde{J}_{||} S_- \\ \tilde{J}_{||} S_+ & 0 & -\tilde{J}_z S_z & \tilde{J}_{12} S_z \\ 0 & -\tilde{J}_{||} S_+ & \tilde{J}_{12} S_z & -\tilde{J}_z S_z \end{pmatrix},$$

$$\tilde{J}_z = J_z \varphi_\uparrow^+(Z)\varphi_\uparrow(Z) = J_z \varphi_\downarrow^+(Z)\varphi_\downarrow(Z),$$
$$\tilde{J}_{12} = J_z \varphi_\uparrow^+(Z)\chi_\uparrow(Z) = J_z \varphi_\downarrow^+(Z)\chi_\downarrow(Z),$$
$$\tilde{J}_{||} = J_{||}\varphi_\uparrow^+(Z)\varphi_\downarrow(Z). \tag{8.13}$$

Hamiltonian (8.13) presents the effective surface model (ESM) for low-energy electrons interacting with a localized spin. The matrix elements in the ESM depend on Z, the magnetic impurity position in the direction normal to surfaces.

In order to better understand the behavior of Z-dependent exchange parameters it is instructive to look at the interaction matrix in the top-bottom basis given in Eq. (7.39), Chapter 7. Below, top (t) and bottom (b) basis states are taken in the sequence: $b\uparrow, t\uparrow, b\downarrow, t\downarrow$:

$$W_{tb}(Z)=\begin{pmatrix} (\tilde{J}_z+\tilde{J}_{12})S_z & 0 & 0 & -\tilde{J}_{||}S_- \\ 0 & (\tilde{J}_z-\tilde{J}_{12})S_z & -\tilde{J}_{||}S_- & 0 \\ 0 & -\tilde{J}_{||}S_+ & -(\tilde{J}_z+\tilde{J}_{12})S_z & 0 \\ -\tilde{J}_{||}S_+ & 0 & 0 & (\tilde{J}_z-\tilde{J}_{12})S_z \end{pmatrix} \tag{8.14}$$

Parameters $\tilde{J}_z+\tilde{J}_{12}$ and $\tilde{J}_z-\tilde{J}_{12}$ describe non-spin-flip *s-d* interaction on bottom and top surfaces, respectively. As shown in Fig. 8.1, they fade out when the impurity moves away from the surface. The spin-flip coupling constant $\tilde{J}_{||}$ is illustrated in Fig. 8.2.

The spin-flip constant $\tilde{J}_{||}$ is proportional to the overlap of top-bottom wave functions and tends to zero in the vicinity of either of two surfaces, it increases between them, being much smaller than near-surface diagonal *s-d* parameters. The *s-d* exchange parameters vs. *z*-position of a magnetic impurity were obtained for a slab thickness of four quintuple layers. For a thinner slab, the spin-flip constant $\tilde{J}_{||}$ increases and in the middle of the slab it becomes only 10 times smaller than the diagonal non-spin-flip parameters.

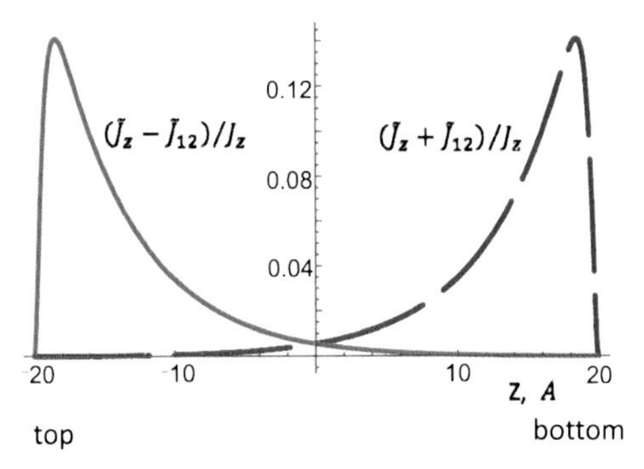

Figure 8.1 *s–d* exchange parameters *vs.* *z*-position of a magnetic impurity.

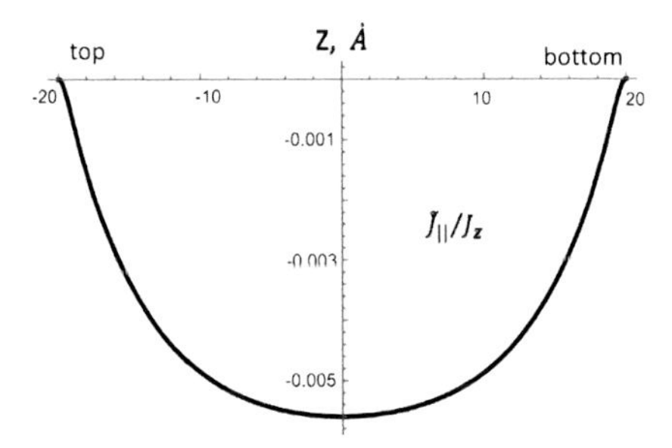

Figure 8.2 The spin-flip exchange parameter vs. the *z*-position of a magnetic impurity.

8.2 Indirect Exchange Interaction Mediated by Surface Electrons

We deal with the Hamiltonian (8.8) where the *s–d* interaction part is given by Eq. (8.12). As shown in Chapter 7, diagonalization of the second term in Eq. (8.8) gives an experimentally observable energy spectrum of surface electrons:

$$H_S^0 + H_S^1 \rightarrow \tilde{H}_0 = \mathrm{diag}(E_{v\uparrow}, E_{c\uparrow}, Ev_{\downarrow}, E_{c\downarrow}),$$

$$E_{c,v\uparrow}(k) = E_0 + \tilde{V}_S + Dk^2 \pm \sqrt{\left(\frac{\Delta_S}{2} - Bk^2\right)^2 + (\tilde{A}_2 k - \tilde{V}_{as})^2},$$

$$E_{c,v\downarrow}(k) = E_0 + \tilde{V}_S + Dk^2 \pm \sqrt{\left(\frac{\Delta_S}{2} - Bk^2\right)^2 + (\tilde{A}_2 k + \tilde{V}_{as})^2}, \quad (8.15)$$

Linear spin splitting near the Γ-point $E_{\uparrow\downarrow} = \pm\alpha_R k$ determines the Rashba parameter

$$a_R = 2\tilde{A}_2 V,\ V = \frac{\tilde{V}_{as}}{\sqrt{\Delta_S^2 + 4\tilde{V}_{as}^2}}. \tag{8.16}$$

The term, proportional to D in Eq. (8.15), makes the effective masses of electrons and holes different. It has a numerically small effect on the indirect exchange interaction and will be neglected.

Transition to the representation which diagonalizes the free-electron Hamiltonian also transforms s-d interaction (8.12). Eigenvectors (7.36) (see Chapter 7) diagonalize H_S^1 and deliver spectrum (8.15). So, the s-d interaction in this representation consists of matrix elements of H_S^{sd} Eq. (8.12) taken with eigenvectors Eq. (7.36) (Chapter 7):

$$H_S^{sd} = \frac{1}{nA}\sum_{ki,k'j} Q_{ij}(Z)\exp(i(\mathbf{k}-\mathbf{k}')\mathbf{R}_{0\|})a_{i\mathbf{k}}^{+}a_{j\mathbf{k}}, \tag{8.17}$$

where

$$Q_{ij}(Z) = \langle i|W(Z)|j\rangle,\ i,j = 1(V\uparrow), 2(C\uparrow), 3(V\downarrow), 4(C\downarrow). \tag{8.18}$$

Matrix Q_{ij} describes the s–d interaction of observable surface electrons with a magnetic atom. The interaction depends on the energy of the incoming electron as well as on the position of the localized spin with respect to both surfaces. We use the approximate Q_{ij} matrix by taking it at $k \to 0$, corresponding to low-energy electrons:

$$\begin{aligned}
Q_{11} &= Q_{44} = -Q_{22} = -Q_{33} = g_S\tilde{J}_{\|}[\mathbf{S}\times\mathbf{n}]_z,\\
Q_{21} &= Q_{12} = -Q_{43} = -Q_{34} = 2\tilde{J}_{\|}|V|[\mathbf{S}\times\mathbf{n}]_z,\\
Q_{13} &= Q_{31}^{*} = S_z(2\tilde{J}_{12}V - \tilde{J}_z) - ig_S\tilde{J}_{\|}(\mathbf{Sn}),\\
Q_{14} &= Q_{41}^{*} = Q_{23} = Q_{32}^{*} = \mathrm{sgn}[V](-\tilde{J}_{12}g_S S_z - 2iV\tilde{J}_{\|}(\mathbf{Sn})),\\
Q_{24} &= Q_{42}^{*} = -S_z(2\tilde{J}_{12}V + J_z) + ig_S\tilde{J}_{\|}(\mathbf{Sn}),\\
g_S &= \Delta_S(\Delta_S^2 + 4\tilde{V}_{as}^2)^{-1/2},\ \mathbf{n} = \mathbf{k}/k.
\end{aligned} \tag{8.19}$$

The spin structure of the interaction is determined by vector and scalar products of impurity spin and the unit vector along the electron in-plane momentum.

Interaction (8.17) generates the indirect exchange between two magnetic atoms separated by the vector $\mathbf{R}_{||} = \mathbf{R}_{||1} - \mathbf{R}_{||2}$ (see Chapter 5):

$$H_{\text{int}} = \frac{T}{2(2\pi)^4 n^2} \sum_{\omega_n} \int\int d\mathbf{k}\, d\mathbf{k}'$$
$$\exp[i\mathbf{R}_{||}(\mathbf{k}-\mathbf{k}')]Tr\{Q(Z_1, \mathbf{n})G(\mathbf{k}, \omega_n)Q(Z_2, \mathbf{n}')G(\mathbf{k}', \omega_n)\}, \quad (8.20)$$

where $G(\mathbf{k}, \omega_n)$ is the Green function,

$$\begin{aligned} G(\mathbf{k}, \omega_n) &= (i\omega_n - \tilde{H}_0 + \mu)^{-1} \\ &= \text{diag}[(i\omega_n - E_{v\uparrow} + \mu)^{-1}, (i\omega_n - E_{c\uparrow} + \mu)^{-1}, \\ &(i\omega_n - E_{v\downarrow} + \mu)^{-1}, (i\omega_n - E_{c\downarrow} + \mu)^{-1}]. \end{aligned} \quad (8.21)$$

μ is the chemical potential, $\omega_n = (2n + 1)\pi T$ is the Matsubara frequency, T is the temperature in energy units. The trace runs over four quantum numbers: bands and spins. After frequency summation in Eq. (8.21), the nonzero contributions at $T = 0$ come from the terms in which the product of the two Green functions comprises the one of initial and the other of the final electron state with energies on both sides of μ.

After trace calculation the structure of Eq. (8.20) depends on the position of the chemical potential. If the chemical potential lies in the energy gap of the surface spectrum, $\mu < \sqrt{\Delta_S^2/4 + V_{as}^2}$, there are no carriers in the surface bands at $T = 0$, and the indirect exchange stems from interband excitations across the gap:

$$\begin{aligned} H_{cv} = &\frac{T}{2(2\pi)^4 n^2} \sum_{\omega_n} \int\int d\mathbf{k}\, d\mathbf{k}' \exp[i\mathbf{R}_{||}(\mathbf{k}-\mathbf{k}')] \times \\ &\{8V^2 M(Z_1, Z_2)[G_{v\uparrow}(\mathbf{k}, \omega_n)G_{c\uparrow}(\mathbf{k}', \omega_n) \\ &+ G_{v\downarrow}(\mathbf{k}, \omega_n)G_{c\downarrow}(\mathbf{k}', \omega_n)][\mathbf{S}_1 \times \mathbf{n}]_z[\mathbf{S}_2 \times \mathbf{n}']_z \\ &+ 8V^2 M(Z_1, Z_2)[G_{c\uparrow}(\mathbf{k}, \omega_n)\, G_{v\downarrow}(\mathbf{k}', \omega_n) \\ &+ G_{v\uparrow}(\mathbf{k}, \omega_n)G_{c\downarrow}(\mathbf{k}', \omega_n](\mathbf{S}_1\mathbf{n})(\mathbf{S}_2\mathbf{n}') \\ &+ 2g_S^2 K(Z_1, Z_2)[G_{c\uparrow}(\mathbf{k}, \omega_n)\, G_{v\downarrow}(\mathbf{k}', \omega_n) \\ &+ G_{v\uparrow}(\mathbf{k}, \omega_n)G_{c\downarrow}(\mathbf{k}', \omega_n)]S_{1z}S_{2z}\}, \\ &M(Z_1, Z_2) = \tilde{J}_{||}(Z_1)\tilde{J}_{||}(Z_2),\ K(Z_1, Z_2) = \tilde{J}_{12}(Z_1)\tilde{J}_{12}(Z_2). \end{aligned} \quad (8.22)$$

In a degenerate slab ($\mu > \sqrt{\Delta_s^2/4 + \widetilde{V}_{as}^2}$), the leading interaction term is of the RKKY-type which originates from excitations around the Fermi energy in the surface conduction band $\varepsilon_F = \mu - \sqrt{\Delta_s^2/4 + \widetilde{V}_{as}^2}$. Conduction Green functions contribute to the exchange of excitations around the Fermi energy, so the intraband indirect exchange is expressed as

$$H_{cc} = \frac{T}{2(2\pi)^4 n^2}\sum_{\omega_n}\int\int d\mathbf{k}\, d\mathbf{k}' \exp[i\mathbf{R}_{\|}(\mathbf{k}-\mathbf{k}')] \times$$
$$\{g_S^2 M(Z_1, Z_2)[G_{c\uparrow}(\mathbf{k},\omega_n)G_{c\uparrow}(\mathbf{k}',\omega_n)$$
$$+ G_{c\downarrow}(\mathbf{k},\omega_n)G_{c\downarrow}(\mathbf{k}',\omega_n)]\{[\mathbf{S}_1\times\mathbf{n}]_z[\mathbf{S}_2\times\mathbf{n}']_z$$
$$+ 2g_S^2 M(Z_1, Z_2)G_{c\uparrow}(\mathbf{k},\omega_n)G_{c\downarrow}(\mathbf{k}',\omega_n)(\mathbf{S}_1\mathbf{n})(\mathbf{S}_2\mathbf{n}')$$
$$+ 2L(Z_1, Z_2)G_{c\uparrow}(\mathbf{k},\omega_n)G_{c\uparrow}(\mathbf{k}',\omega_n)S_{1z}S_{2z}\}, \tag{8.23}$$

$$L(Z_1, Z_2) = \tilde{J}_z(Z_1)\tilde{J}_z(Z_2) + 2V[\tilde{J}_{12}(Z_1)\tilde{J}_z(Z_2)$$
$$+ \tilde{J}_{12}(Z_2)\tilde{J}_z(Z_1)] + 4V^2K(Z_1, Z_2). \tag{8.24}$$

To find out the spin texture of the indirect exchange we perform integration over angles in Eqs. (8.22) and (8.23):

$$I_{1\varphi} = S_{1z}S_{2z}\int \exp[iR_{\|}(\mathbf{k}-\mathbf{k}')]d\varphi_1 d\varphi_2$$
$$= \int_0^{2\pi} d\varphi_1 \exp[ikR_{\|}\cos(\varphi_1)]\int_0^{2\pi} d\varphi_2 \exp[-ik'R_{\|}\cos(\varphi_2)]$$
$$= 4\pi^2 J_0(kR_{\|})J_0(k'R_{\|})S_{1z}S_{2z}, \tag{8.25}$$

$$I_{2\varphi} = \int (\mathbf{S}_1\mathbf{n})(\mathbf{S}_2\mathbf{n}')\exp[i\mathbf{R}_{\|}(\mathbf{k}-\mathbf{k}')]d\varphi_1 d\varphi_2$$
$$= S_1 S_2 \int_0^{2\pi} d\varphi_1 \cos(\alpha_1)\exp[ikR_{\|}\cos(\varphi_1)]$$
$$\times \int_0^{2\pi} d\varphi_2 \cos(\alpha_2)\exp[-ik'R_{\|}\cos(\varphi_2)]$$
$$= \frac{4\pi^2}{R_{\|}^2} J_1(kR_{\|})J_1(k'R_{\|})(\mathbf{S}_1\mathbf{R}_{\|})(\mathbf{S}_2\mathbf{R}_{\|}), \tag{8.26}$$

$$I_{3\varphi} = \int [\mathbf{S}_1\times\mathbf{n}]_z[\mathbf{S}_2\times\mathbf{n}']_z \exp[i\mathbf{R}_{\|}(\mathbf{k}-\mathbf{k}')]d\varphi_1 d\varphi_2$$
$$= \frac{4\pi_2}{R_{\|}^2} J_1(kR_{\|})J_1(k'R)[\mathbf{S}_1\times\mathbf{R}_{\|}]_z[\mathbf{S}_2\times\mathbf{R}_{\|}]_z, \tag{8.27}$$

where $J_{0,1}(x)$ are the Bessel function of the first kind. Calculations were performed making use $\alpha_{1,2}$ expressed in terms of angles $\gamma_{1,2}$ between spin vectors and the vector $\boldsymbol{R}_{||}$ that connects them, as illustrated in Fig. 8.3:

$$\cos(\alpha_1) = \cos(\gamma_1)\cos(\varphi_1) - \sin(\gamma_1)\sin(\varphi_1)$$
$$\cos(\alpha_2) = \cos(\gamma_2)\cos(\varphi_2) - \sin(\gamma_2)\sin(\varphi_2) \tag{8.28}$$

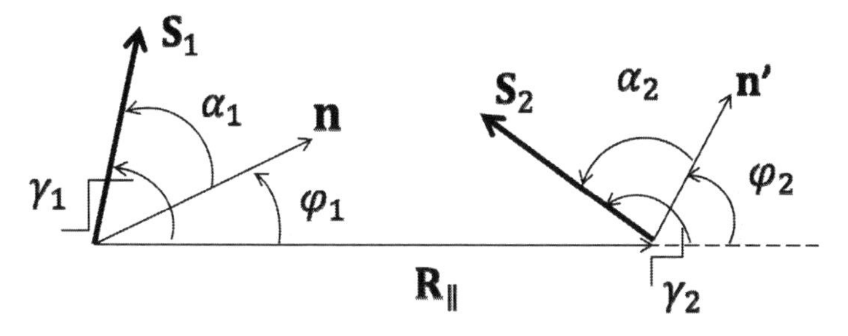

Figure 8.3 Angles in integrals (8.26) and (8.27).

After integration over these angles the interband indirect exchange from Eq. (8.22) can be written as

$$H_{cv} = \frac{1}{4\pi^2 n^2} \times \left\{ \begin{aligned} &\frac{4}{R_{||}^2} V^2 M(Z_1, Z_2)[F^1_{v_\uparrow c_\uparrow} + F^1_{v_\downarrow c_\downarrow}]\,[\mathbf{S}_1 \times \mathbf{R}_{||}]_z [\mathbf{S}_2 \times \mathbf{R}_{||}]_z \\ &+ \frac{4}{R_{||}^2} V^2 M(Z_1, Z_2)\,[F^1_{c_\uparrow v_\downarrow} + F^1_{v_\uparrow c_\downarrow}]\,(\mathbf{S}_1\mathbf{R}_{||})\,(\mathbf{S}_1\mathbf{R}_{||}) \\ &+ g_S^2 K(Z_1, Z_2)[F^0_{c_\uparrow v_\downarrow} + F^0_{v_\uparrow c_\downarrow}] S_{1z} S_{2z}, \end{aligned} \right\} \tag{8.29}$$

where

$$F^{0,1}_{isjs'} = T\sum_{\omega_n} G^{0,1}_{is}(\omega_n, R_{||}) G^{0,1}_{js'}(\omega_n, R_{||}),$$

$$G^{0,1}_{is}(\omega_n, R_{||}) = \int_0^\infty kdk\, J_{0,1}(kR_{||})(i\omega_n - E_{is}(k) + \mu)^{-1} \tag{8.30}$$

Intraband interaction (8.23) has the form:

$$H_{cc} = \frac{1}{8\pi^2 n^2}\left\{\frac{1}{R_{||}^2} g_S^2 M(Z_1, Z_2)[F^1_{c\uparrow c\uparrow} + F^1_{c\downarrow c\downarrow}][\mathbf{S}_1 \times \mathbf{R}_{||}]_z[\mathbf{S}_2 \times \mathbf{R}_{||}]_z \right.$$
$$+ \frac{2}{R_{||}^2} g_S^2 M(Z_1, Z_2) F^1_{c\uparrow c\downarrow}(\mathbf{S}_1\mathbf{R}_{||})(\mathbf{S}_2\mathbf{R}_{||})$$
$$\left. + 2L(Z_1, Z_2) F^0_{c\uparrow c\downarrow} S_{1z} S_{2z} \right\} \qquad (8.31)$$

Some conclusions on possible magnetic phases can be drawn even without actual calculation of the range functions. Coefficients K, M and L carry the dependence of the indirect exchange on positions of magnetic impurities along the z-direction.

The coefficient $M = \tilde{J}_{||}(Z_1)\tilde{J}_{||}(Z_2)$ is the product of two factors each of which tends to zero in the vicinity of either of the surfaces as shown in Fig. 8.2. So, if magnetic atoms are located on either of two surfaces, the terms proportional to M are negligible and the interaction has the Ising-type spin texture favoring magnetic ordering perpendicular to the film surfaces (last terms in Eqs. (8.29) and (8.31)). M-terms come into play if the magnetic atoms are close to the middle of the film in the z-direction, and magnetic ordering caused by M-terms has a complicated spin pattern depending on angles between the vector connecting two impurities and their spin vectors.

For magnetic atoms located at the surfaces the interband indirect exchange (8.29) is fully determined by the K-term whose sign depends on the mutual positions of interacting magnetic atoms. The function $J_{12}(Z)$ that enters the coefficient $K = \tilde{J}_{12}(Z_1)\tilde{J}_{12}(Z_2)$, is illustrated in Fig. 8.4.

It follows from the spatial dependence shown in Fig. 8.4 that the sign of the K-term is positive if interacting spins belong to the same surface and negative if they lay near opposite surfaces. Away from the surfaces, the Ising term in (8.29) tends to zero and a weak interspin coupling is described by the first and second terms. So, the gate-bias-induced complex planar spin texture described by non-Ising terms appears in the middle of the slab. As the positions of spins move toward the surfaces, the non-Ising terms fade out.

In intraband interaction (8.31) the Ising term is governed by the coefficient L which is switchable by an external voltage applied across the film. The voltage dependence of the interaction is illustrated in Fig. 8.5. The numerical example implies that one of the impurity spins is located close to the top of the slab ($Z_1 = -L/2$) and $J_z = J_{||} \equiv J$.

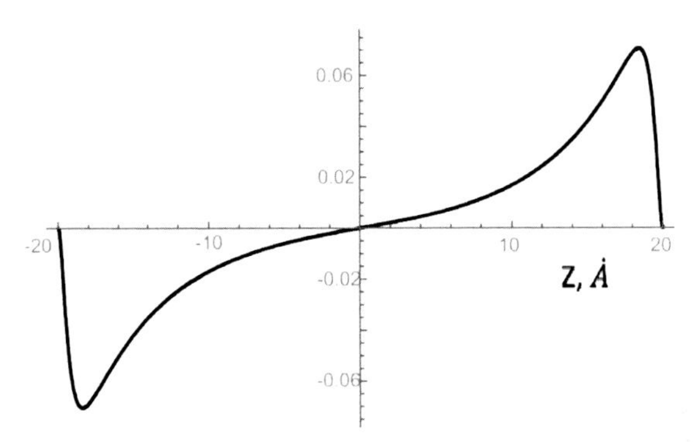

Figure 8.4 Interband exchange parameter $\tilde{J}_{12}(Z)/J_z$ vs. Z-position of a magnetic impurity in the slab.

If the applied voltage changes its sign, the maximum interaction amplitude is switchable between opposite surfaces as shown in Fig. 8.5. For a small bias ($|V_{as}| << \Delta_S$), the graph in Fig. 8.5a becomes symmetrical no matter which surface the two impurities are placed on. The indirect exchange depends only on the distance between them (controlled by the range function only). A large bias leaves nonzero exchange interaction on one surface only, as determined by the sign of the voltage. Thus, by increasing the voltage one can purposely differentiate the surface with nonzero magnetic interaction from one where the interaction is negligibly small.

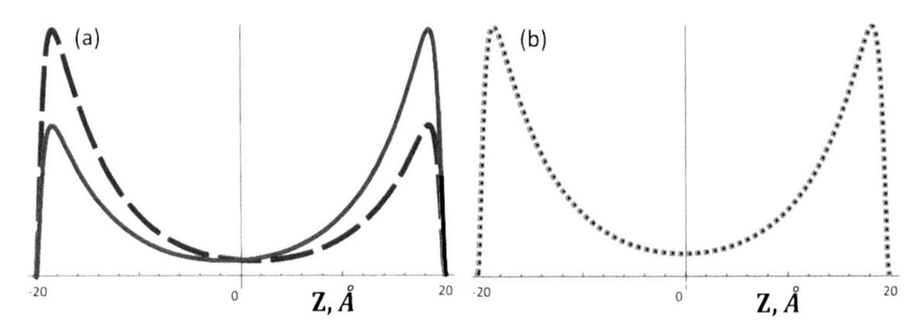

Figure 8.5 Spatial dependence of $L(Z_1, Z_2)/J$ under applied voltage. (a) Solid line: $V_g = 0.35$ V, dashed line: $V_g = -0.35$ V, (b) dotted line: $V = 0$.

8.3 Range Function in Topological Insulator

Since it is not easy to pin the chemical potential to the middle of the surface energy gap, the most probable experimental scenario is described by Eq. (8.31) and corresponds to a doped structure with a finite Fermi level in the surface conduction band. For magnetic atoms placed in the vicinity of surfaces, terms proportional to M are negligible and the interaction is expressed as

$$H_{cc} = V(\mathbf{R})S_{1z}S_{2z},\ V(\mathbf{R}) = \frac{L(Z_1, Z_2)}{4\pi^2 n^2} T\sum_{\omega_n} I_{c\uparrow}(\omega_n)I_{c\downarrow}(\omega_n),$$

$$I_{c\uparrow\downarrow}(\omega_n) = \int_0^{\infty} \frac{kdk\, J_0(kR_{||})}{i\omega_n - E_{c\uparrow\downarrow}(k) + \mu}. \tag{8.32}$$

In order to calculate the range function $V(\mathbf{R})$ analytically, we use the conduction band energy spectrum expanded up to the k^2 term (reference energy is in the middle of the surface energy gap):

$$E_{c,\uparrow\downarrow}(k) = \frac{1}{2}\sqrt{\Delta_S^2 + 4\widetilde{V}_{as}^2} + \widetilde{B}k^2 \pm \alpha_R k,$$

$$\widetilde{B} = D - Bg_S(g_S^2 + 4V^2) + \frac{g_S^2\,\widetilde{A}_2^2}{\sqrt{\Delta_S^2 + 4\widetilde{V}_{as}^2}}\frac{\hbar^2}{2m}. \tag{8.33}$$

In what follows we suppose that $\widetilde{B}$ (inverse effective mass) is positive and then the spectrum as a function of momentum comprises two parabolas shifted with respect to each other by Rashba spin splitting ($\alpha_R \neq 0$).

Using analytical continuation of Bessel functions to a negative k-half-plane

$$J_0(kr) = \frac{1}{2}(H_0^1(kr) + H_0^{(2)}(kr)).$$

$$H_0^{(1)}(ze^{i\pi}) = -H_0^{(2)}(z), \tag{8.34}$$

we obtain the integral over momentum in the form:

$$I_{c\uparrow}(\omega_n) = \frac{1}{2}\int_{-\infty}^{\infty} \frac{kH_0^{(1)}(kR_{||})dk}{i\omega_n - \widetilde{B}k^2 - \alpha_R k + \varepsilon_F}, \tag{8.35}$$

Expression (8.35) accounts for the spin-up band only, and to keep the integration over this band we have to flip $\alpha_R \to -\alpha_R$ in the course of transition to a negative k half-axis. Mere reflection $k \to -k$ would correspond to transition to the spin-down band.

Hankel function $H_0^{(1)}(z)$ is analytical in the upper k-half-plane and tends to zero on the arc $z \to \infty$. So, in order to calculate (8.35) we close the integration path in the upper half-plane. At this point it is important to make sure that the poles of the integrand are located on both sides of the real k-axis. Otherwise the integral would be zero as the path can be closed around the region where the integrand is an analytical function. Moreover, as we will use (8.35) in subsequent summation over Matsubara frequency ω, it is important that appropriate pole positions hold for $\omega > 0$ and $\omega < 0$. The poles

$$k_{1,2} = \frac{1}{2\widetilde{B}}\left[-\alpha_R \pm \sqrt{\alpha_R^2 + 4\widetilde{B}(i\omega_n + \varepsilon F)}\right] \tag{8.36}$$

are shown in Fig. 8.6.

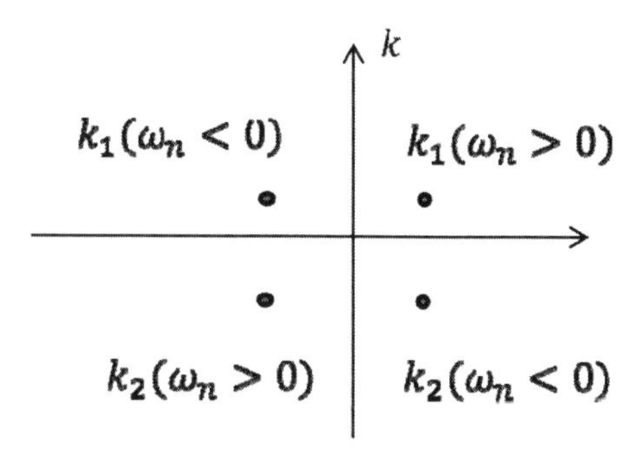

Figure 8.6 Poles of the integrand in (8.35) on the complex k-plane.

Finally, the integral (8.35) is found by calculating the residue

$$I_{c\uparrow}(\omega_n) = -\frac{\pi i}{2\widetilde{B}} \times \frac{-R + \sqrt{R^2 + ix + F}}{\sqrt{R^2 + ix + F}} H_0^{(1)}\left[R_{||}\left(-R + \sqrt{R^2 + ix + F}\right)\right],$$

$$R = \frac{a_R}{2\widetilde{B}};\ F = \frac{\varepsilon_F}{2\widetilde{B}};\ x = \frac{\omega_n}{2\widetilde{B}}. \tag{8.37}$$

Calculating the integral $I_{c\downarrow}(\omega_n)$ in a similar way we obtain

$$I_{c\uparrow}(\omega_n)I_{c\downarrow}(\omega_n) = -\frac{\pi^2}{4\widetilde{B}^2}\left(1 - \frac{R^2}{R^2 + ix + F}\right)H_0^{(1)}\left[R_{||}(-R + \sqrt{R^2 + ix + F})\right]$$
$$\times H_0^{(1)}\left[R_{||}(R + \sqrt{R^2 + ix + F})\right], \tag{8.38}$$

The range function $V(\boldsymbol{R})$, (8.32) contains the sum over Matsubara frequencies and at $T \to 0$ the sum can be replaced with the integral (see Chapter 5):

$$S_M = T\sum_{\omega_n} I_{c\uparrow}(\omega_n)I_{c\downarrow}(\omega_n) = -\frac{i}{2\pi}\int_\Gamma I_{c\uparrow}(\omega)\, I_{c\uparrow}(\omega)d\omega,$$
$$i\omega_n \to \omega + i\, sign(\omega). \tag{8.39}$$

Contour $\Gamma = \Gamma_1 + \Gamma_2$, shown in Fig. 8.7, avoids the branch point $x_0 = -R^2 - F$ and the cut line determined by the square root in Eq. (8.38).

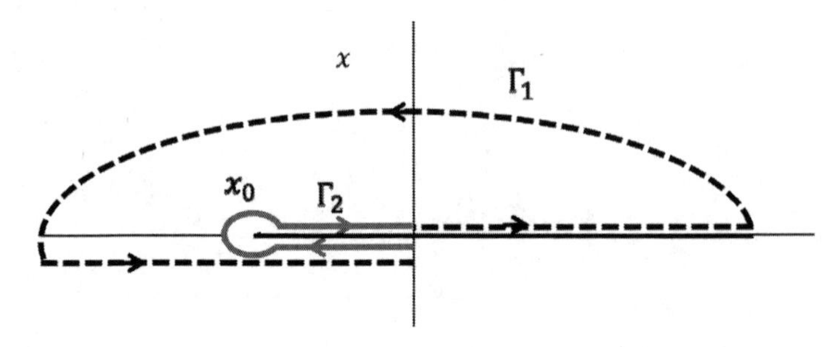

Figure 8.7 Integration contour in Eq. (8.39).

The integral over the circle around the pole tends to zero when the radius decreases, and S_M can be expressed via integration over the upper and lower sides of the path Γ_2:

$$S_M = \frac{i\pi}{4\widetilde{B}}\int_0^{\sqrt{R^2+F}} ydy\left(1 - \frac{R^2}{y^2}\right)\{H_0^{(1)}[R_{||}(-R+y)]H_0^{(1)}[R_{||}(R+y)]$$
$$-H_0^{(1)}[R_{||}(-R-y)]H_0^{(1)}[R_{||}(R-y)]\},$$
$$y = \sqrt{R^2 + x + F}. \tag{8.40}$$

Hankel function $H_0^{(1)}$ has a cut on the negative part of the real axis. To avoid the integration of $H_0^{(1)}$ in this region we use the relation $H_0^{(1)}(-y) = -H_0^{(2)}(y)$, so (8.40) takes the form

$$S_M = \frac{i\pi}{4\tilde{B}} \times$$

$$\left\{\int_0^R yh_1(y)\left(1-\frac{R^2}{y_2}\right)dy + \int_R^{\sqrt{R^2+F}} yh_2(y)\left(1-\frac{R^2}{y^2}\right)\right\}dy,$$

$$h_1(y) = H_0^{(2)}[R_{||}(R+y)]H_0^{(1)}[R_{||}(R-y)] - H_0^{(2)}[R_{||}(R-y)]H_0^{(1)}[R_{||}(R+y)],$$

$$h_2(y) = H_0^{(1)}[R_{||}(y-R)]H_0^{(1)}[R_{||}(y+R)] - H_0^{(2)}[R_{||}(y+R)]H_0^{(2)}[R_{||}(y-R)].$$

(8.41)

Changing the variable to $s = yR_{||}$ and using notations (8.37) we get the final result for the range function:

$$V(R) = -\frac{L(Z_1, Z_2)m}{4\pi n^2 h^2 R_{||}^2}$$

$$\times \left\{\int_0^{R_{||}k_R} s\Phi(s)P(s)ds + \int_{R_{||}k_R}^{R_{||}\sqrt{k_R^2+k_F^2}} s\Phi(s)Q(s)ds\right\},$$

$$\Phi(s) = 1-(R_{||}k_R)^2/s^2; \; k_R = \alpha_R m/\hbar^2; \; k_F = 2m\varepsilon_F/\hbar^2,$$

$$P(s) = J_0(R_{||}k_R+s)N_0(R_{||}k_R-s) - J_0(R_{||}k_R-s)N_0(R_{||}k_R+s)$$

$$Q(s) = J_0(s-R_{||}k_R)N_0(s+R_{||}k_R) + J_0(s+R_{||}k_R)N_0(s-R_{||}k_R)$$

(8.42)

where $N_0(S)$, is the Newmann function (Bessel function of the second kind), k_R and k_F are the Rashba and Fermi momentum, respectively. In Eq. (8.42) we used the identities $H_0^1(s) = J_0(s) + iN_0(s)$, $H_0^2(s) = J_0(s) - iN_0(s)$.

If the bias and then the parameter $k_R \sim V_{as}$ tends to zero, the second term in Eq. (8.42) becomes a standard oscillating 2D-RKKY range function (see Chapter 5 and Refs. [11, 12]). Under an applied voltage the range function contains additional beatings, with a tunable period proportional to $1/k_R$. The result is to be expected from the qualitative standpoint as the beating is a consequence of two distinct Fermi momenta in the spin-split Rashba electron gas. The first integral is an additional contribution to the RKKY range function that presents the signature of Rashba spin splitting. It should be noted that spin-splitting-related features in RKKY may exist even without external bias as films are grown on a substrate and the built-in electric field near broken surfaces makes them non-equivalent, thus violating inversion symmetry.

8.4 Conclusions

In this chapter we consider the indirect exchange interaction mediated by degenerate surface electrons. In the nondegenerate state the chemical potential is placed within the surface energy gap and interband terms (8.29) determine the magnetic ordering in the slab. Based on qualitative considerations one may predict some features of the magnetic interaction in this case.

If $B\Delta_S < 0$ the surface spectrum has a direct gap and the range function should fade out exponentially with $R_{||}$ as it happens in an ordinary non-degenerate semiconductor (see Chapter 5).

In the topologically non-trivial phase, $B\Delta_S > 0$, the spectrum is inverted (see Fig. 7.8, Chapter 7). Two types of virtual electron-hole transitions contributing to the indirect exchange become possible: 1) vertical transitions across the minimum gaps at $\pm k_0$, $k_0 = \sqrt{(\Delta_S/2B)}$, and 2) transitions across the gap between minima at different momenta with momentum transfer $K \approx 2k_0$. The range function oscillates with a period proportional to K^{-1} and exponentially decaying amplitude $\sim\exp(-R_{||}/r_0)$, where $r_0 \approx B^{3/4}(\tilde{A}_2)^{-1/2}(2\Delta_S)^{-1/4}$ if $V_{AS} = 0$ (see Problem 1). A similar type of range function appears in indirect-gap semiconductors [13], graphene with the energy gap induced by spin-orbit interaction [14], and excitonic insulators [15]. So, the signature of the non-degenerate topological phase is oscillating indirect exchange with exponentially decaying amplitude whereas in a trivial insulator the non-degenerate surface states would mediate a monotonically decreasing exponential range function.

In conclusion, the low-energy effective *s-d* interaction model in TI has been developed by projecting the bulk *s-d* interaction onto surface states. It is shown that magnetic atoms interact with the surface electrons through position-sensitive *s-d* interaction that can be controlled by gate bias. The range function of the indirect exchange oscillates in a degenerate sample and, in addition to main oscillations determined by a finite Fermi momentum, it acquires zero magnetic field beatings with a period related to the magnitude of the Rashba spin splitting.

Problems

8.1 Estimate the period of oscillations and the decrement of exponential decay of the range function mediated by non-degenerate surface electrons. See Section 8.4 and Fig. 7.8 of Chapter 7.

8.2 Estimate the gate voltage which turns off exponential distance dependence in the range function mentioned in Problem 8.1.

References

1. Henk J, Flieger M, Maznichenko IV, Mertig I, Ernst A, Eremeev SV, Chulkov EV (2012) Topological character and magnetism of the Dirac state in Mn-Doped Bi_2Te_3, *Phys Rev Lett,* **109**, 076801.
2. Zhang J, Chang C-Z, Tang P, Zhang Z, Feng X, Li K, Wang L-l, Chen X, Liu C, Duan W, He K, Xue Q-K, Ma X, Wang Y (2013) Topology-driven magnetic phase transition in topological insulators, *Science,* **339**, 1582–1586.
3. Liu Q, Liu C-X, Xu C, Qi X-L, Zhang S-C (2009) Magnetic impurities on the surface of a topological insulator, *Phys Rev Lett,* **102**, 156603.
4. Zhu J-J, Yao DX, Zhang S-C, Chang K (2011) Electrically controllable surface magnetism on the surface of topological insulators, *Phys Rev Lett,* **106**, 097201.
5. Abanin DA, Pesin DA (2011) Ordering of magnetic impurities and tunable electronic properties of topological insulators, *Phys Rev Lett,* **106**, 136802.
6. Checkelsky JG, Ye J, Onose Y, Iwasa Y, Tokura Y (2012) Dirac-fermion-mediated ferromagnetism in a topological insulator, *Nat Phys,* **8**, 729–733.
7. Ye F, Ding GH, Zhai H, Su ZB (2011) Spin helix of magnetic impurities in two-dimensional helical metal, *Europhys Lett,* **90**, 47001.
8. Sun J, Chen L, Lin H-Q (2014) Spin-spin interaction in the bulk of topological insulators, *Phys Rev B,* **89**, 115101.
9. Zhu Z-H, Levy G, Ludbrook B, Veenstra CN, Rosen JA, Comin R, Wong D, Dosanjh P, Ubaldini A, Syers P, Butch NP, Paglione J, Efimov IS, Damascelli A (2011) Rashba spin-splitting control at the surface of the topological insulator Bi_2Se_3, *Phys Rev Lett,* **107**, 186405.
10. Litvinov VI (2014) Magnetic exchange interaction in topological insulators, *Phys Rev B,* **89**, 235316.

11. Fischer B, Klein MW (1975) Magnetic and nonmagnetic impurities in two-dimensional metals, *Phys Rev B*, **11**, 2025–2029.
12. Litvinov VI, Dugaev VK (1996) RKKY interaction in one- and two-dimensional electron gases, *Phys Rev B*, **58**(7), 3584–3585.
13. Abrikosov AA (1980) Spin-glass with a semiconductor as host, *J Low Temp Phys*, **39**(1/2), 217–229.
14. Dugaev VK, Litvinov VI, Barnas J (2006) Exchange interaction of magnetic impurities in graphene, *Phys Rev B*, **74**, 224438.
15. Litvinov VI (1985) Spin glass in excitonic insulator, *Sov Phys Solid State*, **27**(4), 740–741.

Index